T0223697

SpringerBriefs in Education

More information about this series at http://www.springer.com/series/8914

Yew-Jin Lee · Mijung Kim
Qingna Jin · Hye-Gyoung Yoon
Kenji Matsubara

East-Asian Primary Science Curricula

An Overview Using Revised Bloom's Taxonomy

 Springer

Yew-Jin Lee
National Institute of Education
Nanyang Technological University
Singapore
Singapore

Hye-Gyoung Yoon
Chuncheon National University of
  Education
Chuncheon, Kangwon-do
Korea (Republic of)

Mijung Kim
University of Alberta
Edmonton, AB
Canada

Kenji Matsubara
National Institute for Educational Policy
  Research
Tokyo
Japan

Qingna Jin
University of Alberta
Edmonton, AB
Canada

ISSN 2211-1921   ISSN 2211-193X   (electronic)
SpringerBriefs in Education
ISBN 978-981-10-2689-8   ISBN 978-981-10-2690-4   (eBook)
DOI 10.1007/978-981-10-2690-4

Library of Congress Control Number: 2016952514

Printed on acid-free paper

This Springer imprint is published by Springer Nature
The registered company is Springer Nature Singapore Pte Ltd.
The registered company address is: 152 Beach Road, #22-06/08 Gateway East, Singapore 189721, Singapore

Acknowledgements

We thank Dr. Subramaniam Ramanathan from the National Institute of Education (Singapore) for valuable advice and generous help on the interrater reliability calculations.

Contents

List of Figures

List of Tables

List of Appendices

Chapter 1
Primary Science Curricula: Past and Present Realities

The Unknown Intellectual Demands of Science Curricula

That science has assumed a nigh indispensable and ubiquitous place within primary schools around the world is a fact that few educators would deny. Reasons for its contemporary currency are plentiful and range from early preparation for life and work in an increasingly technological world to exposure to one of humanity's greatest cultural treasures. Given that not all young people might pursue the study of science beyond elementary school due to poverty, falling motivation, or lack of access, being able to learn and experience something about science at this stage no matter how modest becomes all the more precious.

These concerns about the nature of scientific content and the spectrum of educative experiences that a child might encounter during this critical period form the central focus of this book. In Lee Shulman's words, such research contributes toward better understanding of what constitutes vertical curriculum knowledge in science, an essential component of expert teaching (Shulman 1986). We are particularly interested in examining the intellectual demands of the intended or planned curriculum in primary science among six jurisdictions in East Asia. Our research thus extends preliminary work from Korea and Singapore that has recently been conducted by three of the authors (Lee et al. 2015). A greater sampling of educational systems in Asia and elsewhere would have been ideal although a lack of resources have curtailed these ambitions.

Nevertheless, such comparative research into an important but strangely neglected field are timely because we describe together for the first time the cognitive processes and knowledge levels of primary science curricula from Hong Kong (SAR), Japan, the Republic of Korea (South Korea), the People's Republic of

© The Author(s) 2017
Y.-J. Lee et al., *East-Asian Primary Science Curricula*,
SpringerBriefs in Education, DOI 10.1007/978-981-10-2690-4_1

China, Taiwan, and Singapore.[1] This in-depth examination of their learning objectives—some of which have just undergone revision (e.g., Korea in 2015, Singapore in 2014)—must furthermore be viewed in the light of another growing policy consequence of globalization—"stating educational standards in terms of student outcomes is a relatively new experience for many countries" (DeBoer 2011, p. 571). And help us to achieve our goals, we employ one of the most widely recognized and adaptable tools in curriculum research—revised Bloom's Taxonomy (RBT) (Anderson et al. 2001).

As acknowledged high-performing states in international achievement tests, such as the Programme for International Student Assessment (PISA) and Trends in International Mathematics and Science Study (TIMMS), our results may be of value to policymakers, teachers, science educators, and comparative education researchers. Given the prominence of these states on these measures, it is hoped that analyzing their intended curricula in this manner can offer international audiences tentative insights into why or how they have succeeded so well in educating young people about science. More significantly, we insist that unless fundamental knowledge about what students are expected to know and to what degree or quality, any critique or impetus for educational reform in these states would be proceeding almost blindly.

The results that we report in this book also has consequences for these governments who wish to provide an equitable education for their future citizens—all youth, especially at-risk or marginalized youth, are obligated to receive instruction in science that is not compromised in intellectual rigor as we explain later. Finally, in the absence of knowledge about the level of intellectual demands in the curriculum, alignment between the intended, taught, and assessed curriculum would be compromised and have negative consequences for general student achievement. While research in all these aforementioned areas might already exist, they have typically been published in local or vernacular outlets, which have very limited readership as well as visibility. Our study therefore promises to be a key contribution in English including our own translations of many learning outcomes hitherto written only in native languages such as those from China, Taiwan, and Korea.

What Our Study Is not and Its Limitations

It is clear that various stakeholders—local and international—would benefit from analyzing our findings that have theoretical and practical significance, including policymakers, researchers, and educators. A strong temptation, however, would be to regard these findings as the latest photo-finish from an intense competition to

---

[1]In the rest of this book, we will refer to Hong Kong (SAR) as Hong Kong, the People's Republic of China as China or mainland China, and the Republic of Korea as Korea. Technically speaking, Hong Kong is a special administrative region (SAR) of China, a territory and not a state although for convenience it will be called as such here. By designating all these six regions or territories as states, we wish to remain neutral as to their legal or political status within international politics.

confirm which state is the "best" in whatever way one would care to define that notorious label. We thoroughly reject such narrow interpretations of the data. Our rationales are instead fueled by curiosity and desire to understand the intended primary science curriculum in all its rich variations, at least among six East-Asian states that have rather similar philosophies concerning teaching and learning. Unique strengths and shortcomings from each state might certainly become apparent when placed side by side as we have done, but we wish to remind readers that there are many good reasons for these variations. Cultural, geographical, political, philosophical, and historical differences have always mediated the formation of any curriculum in an interlocking manner. We too invite readers to recognize how one's personal gaze and ideas of what is good, beautiful, or desirable in education are inherently motivated, coming from a certain standpoint depending on one's background and experiences. So while this book does indeed perform the task of comparison, at no time does it attempt to judge. We are hopeful that the findings will be employed in the same vein of refined connoisseurship as how many comparative researchers have described their labors—to understand oneself or one's context better through a vigorous process of appreciating the other (Manzon 2014).

In addition, our research samples only a thin slice in time following a particular conception of what constitutes intellectual demand according to RBT; curricula are objects that evolve, sometimes with great speed and in an unpredictable manner. For example, historical analyses from Iceland showed major decadal transformations in the ideologies underlying their intended science curriculum (Thorolfsson et al. 2012). Our comparative study should hence be understood with these manifest limitations of time and space. If anything, they suggest that research like ours is forever a context-bound work-in-progress though we are puzzled by the sheer lack of empirical evidence from Asia that prompted our investigations in the first place. Observant readers will also realize that the knowledge levels and cognitive processes (i.e., intellectual or epistemic knowledge) found in RBT bear no resemblances with more recent versions of scientific literacy as espoused by the PISA tests and American *Next Generation Science Standards*. Our data will therefore yield a generic psychological understanding of intellectual demand rather than something specific to science per se just as we are fully aware of the criticisms of systematic, technical-rational modes of policymaking frequently associated with mandatory curriculum objectives (e.g., Bencze and Alsop 2009).

Above all, teaching remains an incredibly complex and multifaceted activity that can proceed in generating meaningful learning despite being led by an uninspiring, tightly scripted curriculum, meager educational provisioning, or even poor subject matter knowledge among instructors (Shechtman et al. 2010). Yet we hear that having scripted lessons plans, and by extension a standardized curriculum within a country can compensate for wide variations in teacher quality within some Asian states (ADB 2015). The experienced curriculum, for the most part, is therefore what occurs when teachers and students engage with subject matter on a daily basis that takes precedence over what has been stated in the planned curriculum, textbooks, or other official policy documents. We are indeed forced to be modest in our desire to better understand how and what potential opportunities to learn are contained within the intended primary science curriculum in the light of these subtle albeit consequential mediators of learning. We know that teachers, teaching practices, textbooks, and various curriculum documents are all major facilitators of learning with differing effects on student achievement (see Hiebert and Grouws 2007).

The following sections revisit some of our intentions and claims of our book in greater detail; we describe how primary science curricula have despite various national distinctives or programs experienced a remarkable worldwide convergence over the past four decades. Next, we explain the need for more critical research into arguably one of the least known aspects of primary science curricula—the intellectual demands that they make on learners. Before we present the methods of the coding process and results of our analysis in the next two chapters, we reestablish how this study satisfies a gap in curriculum research that is especially lacking from Asian states.

Overview of Primary Science Education

How has the philosophy and teaching of primary science been transformed over time? What were some of the overarching curricular patterns and influences that have shaped the development of this school subject across the world? Due to globalization, curricula and school systems have become more and more similar worldwide based on the standardization of educational approaches and instruments, especially with regard to performance-based achievement (Jenkins 2015; Loomis et al. 2008; Wilkinson 2013). This robust unifying trend detected more than 20 years ago in education (Benavot et al. 1991) is paralleled too in science education, especially with regard to the disciplinary structure of curriculum, intended goals of scientific literacy, classroom practices, and what is regarded as good science teaching (see Nilsson 2015). Based on a comprehensive coding of 265 primary science textbooks (Grades 3–7) from 60 countries published from around 1900 to 1995, Elizabeth McEneaney (2003) reported the following changes in the emphases in primary science as illustrated in Fig. 1.1.

Compared to primary science instruction in the beginning of the last century, textbooks catering for elementary grades since the 1960s have depicted science as supplying greater personal relevance and with opportunities for personal fun or enjoyment. There were also more frequent depictions of ordinary people (such as young children like the readers themselves) doing exciting scientific activities and initiating research questions; the normally formidable barriers between scientists and laypeople were being downplayed (although not completely so) thereby inviting greater participation and emotional closeness towards science from young learners. By celebrating a greater sense of agency and expertise among diverse groups of learners, nature was likewise increasingly depicted in primary science textbooks as an environment that was benign, ordered, and manageable through human intervention such as theirs. Other researchers and historians of science have similarly reported how science curricula around the world were encouraging the

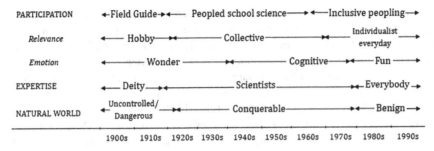

Fig. 1.1 Global trends in the emphases and concerns in primary science as reflected in the coding of content from primary science textbooks across 60 countries. Adapted from McEneaney (2003)

learning of science through hands-on inquiry rather than through rote memorization of isolated facts (Anderson 2002; Atkin and Black 2003; Rudolph 2002). These seemingly dramatic shifts, of course, must be viewed in the light of curriculum-making that was increasingly the purview of large and well-equipped government ministries rather than local school districts or even individual teachers (Elyon 2014; Harlen 2014a). As mentioned, policy borrowing between states was also becoming common at the same time as the prescriptive influences of international testing and competitions exacerbated this movement toward greater accountability and standardization in education. Certainly, we can increasingly detect within many national curriculum documents a prevailing discourse that has emphasized so-called twenty-first century skills for life and work in an uncertain and unpredictable future (Sinnema and Aitkin 2013).

Lack of Research into Primary Science Curricula

To a large extent, the evolution and development of primary science in Asia (see Jenkins 2003, 2015) have run in tandem with these worldwide trends, such as what has happened in Singapore (Chin and Poon 2014), Japan, and the Philippines (Pawilen and Sumida 2005) among many other countries (see Kim, Chu, and Lim (Kim et al. 2015) for a review of secondary science curriculum reforms in East Asia). While it is almost a truism that research into the primary science curriculum is vitally important given these developments, it is unusual that so little has actually been published on this subject apart from the general trends that have been mentioned in the previous section. It is as though this is a trivial question, or that inquiry here has passed into the stage whereby no major insights for theory or practice can be further uncovered.

To some extent, this absence of interest was a direct consequence of an initial skepticism about the ability of learners of primary science to develop an inquiring mind as much as a moral panic about a premature induction into secondary science that saw resistance to the adoption of early years science among a number of

Western countries (Atkin and Black 2003; Harlen 2014a, b). Those in the developing world, on the contrary, were among the earliest to grant their youngest citizens access to school experiences in primary science apparently because of a greater faith in the (often unsubstantiated) benefits of science education that was simultaneously being facilitated by large international development agencies, such as the World Bank, the Asian Development Bank, and UNESCO (United Nations Educational, Scientific, Cultural Organization) (Harlen 1994).

Be that as it may, a large gap in our knowledge still exists within Asian contexts, where a multitude of official languages and dialects, political systems, colonial histories, and issues of income disparity, gender equity, and intractable socioeconomic problems have prevented many states here from achieving universal access to primary education, much less being able to adequately reform or improve the implementation of science curricula inside their borders. This state of affairs where we know so little about what passes for primary science in Asia has therefore moved some policy researchers to declare that "with all the interest in providing an instructionally effective and financially efficient educational environment, it is surprising how little is said (or known) about one of the most important components of schooling in the modern world: the curriculum" (Benavot and Kamens 1989, p. 3).

Of course, access to education in the first instance continues to pose a challenge for many Asian countries where more than half of humanity calls their home and where illiteracy is rampant (ADB 2015; UNHDR 2014). Coupled with large-scale natural disasters and increasing episodes of food shortages due to climate change, it is understandable that many governments in Asia have focused their priorities on more pressing social problems. Yet, one cannot run away from highly visible disparities in education that exist between developing and developed countries around the world. Data from national expenditures on education a decade ago already showed that the United States of America that has 4 % of the global population (between 5 and 25 years of age) accounted for 28 % of the entire world's spending on education (UNESCO 2007). Insofar that the second largest share of public spending on education (18 %) came from East Asia (China & Japan accounted for more than half of total sum here) and the Pacific, it must not be forgotten that at least 17 % of students enrolled in school here were to be located within China alone.

Why Knowing the Intellectual Demands of Curricula Matters

Arguably one of the most critical and useful aspects of research into the curriculum is the cognitive demands that they make on learners. Even among countries that have traditionally advocated child-centered, experiential ways of teaching are now increasingly moving towards curriculum-making determined at the national/ministry level. This truly represents a sea change in educational policymaking and societal expectations among liberal democracies, such as Australia, Denmark, Germany, New Zealand, and the United States just to name a few (OECD 2015). Students and teachers here now have to take reference from a set of public learning objectives or standards to benchmark their rate of progress and testing regimes in numerous school subjects. Acceptance and compliance to these standards continue to be highly contentious and politically charged affairs, which is to be expected (e.g., PDK/Gallop Poll 2015). Yet without knowing what are the cognitive demands of these objectives or standards, assessment, and teaching would naturally be unfocused because of misalignments in the intended, taught, and assessed curriculum. It was this very basic problem that was presumed to be the main cause for the relatively poor performance of students in the United States in mathematics and science—American students were reportedly covering a wide variety of topics at the expense of meaningful learning of concepts (Schmidt et al. 2001). On the other hand, it was alleged that Canadian students were being exposed to lower than expected levels of thinking according to RBT, which thus prevented them from cultivating robust forms of scientific literacy (Fitzpatrick and Schulz 2015).

Another set of reasons for the incontrovertible need to establish what are the cognitive demands within any school curriculum revolves around what is known as "powerful knowledge" (Young 2007). This is an idea taken from research in the sociology of knowledge; rather than school knowledge being regarded as overly abstract and/or alienating for weaker/marginalised pupils (which obliges watering down the syllabus), it is actually in the long-term interests of these students to acquire if not master these theoretical forms of knowledge. This acquisition of difficult-to-learn concepts in school thus amounts to accumulating symbolic/cultural power to enable learners to later become an active member in society's dialogues about itself as how its chief proponent Michael Young put it. Everyday learning experiences such as that obtained from the home are simply unable to transcend or be generalizable in other contexts (e.g., workplaces, society) as compared to theoretical concepts that schools do so well at imparting through the vehicle of the curriculum. In this sociological understanding of powerful knowledge, one accordingly has to ensure that school experiences not only have multiple opportunities to acquire powerful knowledge through teaching (planned or unplanned), but that the intended curriculum itself has consciously incorporated complex, theoretical concepts, that is, powerful knowledge. Educational authorities that assume school curricula would have high intellectual demands by default are oftentimes mistaken as in the previously described situation in Canada. This argument for providing more equitable education through powerful knowledge finds much resonance with our basic call to understand what are the intellectual demands within primary science curricula.

We acknowledge that there is widespread skepticism of the curriculum as document in the developed or Western world. These are legitimate stances although this disdain or lack of respect for understanding the curriculum as program is perhaps unhelpful in the light of our aforementioned arguments; all educators cannot fail to see that educational visions are being translated into classroom realities even if imperfectly realized. The success of schooling as an institution is inexorable and is prior to any critique of education as a normative, social enterprise (Westbury 2008). We have earlier labored the point that most of Asia remains a highly irregular patchwork of wealth and poverty harboring cause for both optimism and despair. Yet, there are some places or programs here that can offer potentially useful insights into teaching and learning for the rest of the world. With a number of possibilities to choose from, our focus is on the intended primary science curriculum from six East-Asian states that have reported to have provided challenging, high-quality instructional opportunities in mathematics and science. By no means are these states free from problems when educating their youth or have perfected outstanding curricula that amply serve the needs of every child. On the contrary, we will only have a clearer picture of why their national institutions, policies, and programs have delivered (or failed) when the former are open for stringent inspection by the international community of scholars: What you see here thus represents our attempt toward this ambitious endeavor.

References

ADB [Asian Development Bank]. (2015). *Key indicators for Asia and the Pacific: 2015 46th Edition*. Manila, Philippines: Asian Development Bank.

Anderson, L. W. (2002). Curricular alignment: A re-examination. *Theory Into Practice, 41*(4), 255–260.

Anderson, L. W., Krathwohl, D. R., Airasian, P. W., Cruikshank, K. A., Mayer, R. E., Pintrich... Wittrock, M. C. (Eds.). (2001). *A taxonomy for learning, teaching and assessing. A revision of Bloom's taxonomy of educational objectives*. White Plains, NY: Addison-Wesley Longman.

Atkin, M. J., & Black, P. (2003). *Inside science education reform: A history of curricular and policy change*. New York, NY: Teachers College Press.

Benavot, A., Cha, Y.-K., Kamens, D., Meyer, J. W., & Wong, S.-Y. (1991). Knowledge for the masses: World models and national curricula, 1920-1986. *American Sociological review, 56*, 85–100.

Benavot, A., & Kamens, D. (1989). *The curricular content of primary education in developing countries*. Washington, DC: The World Bank.

Bencze, J. L., & Alsop, S. (2009). A critical and creative inquiry into school science inquiry. In W.-M. Roth & K. Tobin (Eds.), *The world of science education: North America* (pp. 27–47). Rotterdam: Sense.

Chin, T.-Y., & Poon, C. L. (2014). Design and implementation of the national primary science curriculum: A partnership approach in Singapore. In A.-L. Tan, C. L. Poon, & S. S. L. Lim (Eds.), *Inquiry into the Singapore science classroom: Research and practices* (pp. 27–46). Dordrecht: Springer.

DeBoer, G. E. (2011). The globalization of science education. *Journal of Research in Science Teaching, 48*, 567–591.

Elyon, B.-S. (2014). Curriculum development. In R. Gunstone (Ed.), *Encyclopaedia of science education*. Available http://www.springerreference.com/docs/html/chapterdbid/303000.html

Fitzpatrick, B., & Schulz, H. (2015). Do curriculum outcomes and assessment activities in science encourage higher order thinking? *Canadian Journal of Science, Mathematics and Technology Education, 15*, 136–154.

Harlen, W. (1994). Primary science. In T. Husén & T. N. Postlethwaite (Eds.), *The international encyclopedia of education* (pp. 5328–5335). Oxford: Pergamon.

Harlen, W. (2014a). Primary/elementary school science curriculum. In R. Gunstone (Ed.), *Encyclopaedia of science education*. Available http://www.springerreference.com/docs/html/chapterdbid/303020.html

Harlen, W. (2014b). Primary/elementary school science curriculum projects. In R. Gunstone (Ed.), *Encyclopaedia of science education*. Available http://www.springerreference.com/docs/html/chapterdbid/303021.html

Hiebert, J., & Grouws, D. A. (2007). The effects of classroom mathematics teaching on students' learning. In F. K. Lester (Ed.), *Second handbook of research on mathematics teaching and learning* (pp. 371–404). Greenwich, CT: Information Age.

Jenkins, E. W. (Ed.). (2003). *Innovations in science and technology education* (Vol. VIII). Paris: UNESCO Publishing.

Jenkins, E. W. (2015). Children and the teaching and learning of science: A historical perspective. In K. Schultheis & A. Pfrang (Eds.), *Children's perspective on school teaching and learning: Studies of the educational experience of children* (pp. 143–161). Zurich, Switzerland: LIT Verlag.

Kim, Y., Chu, H.-Y., & Lim, G. (2015). Science curriculum changes and STEM education in East Asia. In M. S. Khine (Ed.), *Science education in East Asia: Pedagogical innovations and research-informed practices* (pp. 149–226). Dordrecht: Springer.

Lee, Y.-J., Kim, M., & Yoon, H.-G. (2015). The intellectual demands of the intended primary science curriculum in Korea and Singapore: An analysis based on revised Bloom's taxonomy. *International Journal of Science Education, 37*, 2193–2213.

Loomis, S., Rodriguez, J., & Tillman, R. (2008). Developing into similarity: Global teacher education in the twenty-first century. *European Journal of Teacher Education, 31*(3), 233–245.

Manzon, M. (2014). Comparing places. In M. Bray, B. Adamson, & M. Mason (Eds.), *Comparative education research: Approaches and methods* (pp. 97–137). Hong Kong: Springer & Comparative Education Research Centre, The University of Hong Kong.

McEneaney, E. H. (2003). Elements of a contemporary primary school science. In G. S. Drori, J. W. Meyer, F. O. Ramirez, & E. Schofer (Eds.), *Science in the modern world polity: Institutionalization and globalization* (pp. 136–154). Stanford, CA: Stanford University Press.

Nilsson, P. (2015). *Primary/elementary science teacher education*. Available from http://www.springerreference.com/docs/html/chapterdbid/303084.html

OECD. (2015). *Education policy outlook 2015: Making reforms happen*. Available http://www.keepeek.com/Digital-Asset-Management/oecd/education/education-policy-outlook-2015_9789264225442-en#page1

Phi Delta Kappan [PDK]/Gallop Poll. (2015). *Testing doesn't measure up for Americans*. Available from http://pdkpoll2015.pdkintl.org/wp-content/uploads/2015/08/pdkpoll47_2015.pdf

Pawilen, G. T., & Sumida, M. (2005). A comparative study of the elementary science curriculum of Philippines and Japan. *Ehime University Faculty of Education Bulletin, 52*(1), 167–180.

Rudolph, J. S. (2002). *Scientists in the classroom: The cold war reconstruction of American science education*. New York, NY: Palgrave.

Schmidt, W., McKnight, C., Houang, R., Wang, H. C., Wiley, D., Cogan, L., et al. (2001). *Why schools matter: A cross-national comparison of curriculum and teaching*. San Francisco, CA: Jossey-Bass.

Shechtman, N., Roschelle, J., Haertel, G., & Knudsen, J. (2010). Investigating links from teacher knowledge, to classroom practice, to student learning in the instructional system of the middle-school mathematics classroom. *Cognition and Instruction, 28*, 317–359.

Shulman, L. S. (1986). Those who understand: Knowledge growth in teaching. *Educational Researcher, 15*, 4–14.

Sinnema, C., & Aitkin, G. (2013). Emerging international trends in curriculum. In M. Priestly & G. Biesta (Eds.), *Reinventing the curriculum: New trends in curriculum policy and practice* (pp. 141–163). London: Bloomsbury Academic.

Thorolfsson, M., Finnbogason, G. E., & Mcdonald, A. (2012). A perspective on the intended science curriculum in Iceland and its 'transformation' over a period of 50 years. *International Journal of Science Education, 34*, 2641–2665.

UNESCO [United Nations Educational, Scientific, Cultural Organization]. (2007). *Global education digest 2007: Comparing education statistics across the world.* Montreal: UNESCO Institute for Statistics.

UNHDR [United Nations Human Development Report]. (2014). *Data.* Available http://hdr.undp.org/en/data

Westbury, I. (2008). Making curricula: Why do states make curricula, and why? In F. M. Connelly, M. F. He, & J. Phillion (Eds.), *The SAGE handbook of curriculum and instruction* (pp. 45–65). Thousand Oaks, CA: Sage Publications.

Wilkinson, G. (2013). McSchools for McWorld? Mediating global pressures with a McDonaldizing education policy response. In G. Ritzer (Ed.), *McDonaldization: The reader* (3rd ed., pp. 149–157). Thousand Oaks, CA: Sage.

Young, M. (2007). *Bringing knowledge back in: From social constructivism to social realism in the sociology of education.* London: Routledge.

Chapter 2
Revised Bloom's Taxonomy—The Swiss Army Knife in Curriculum Research

Tools to Analyze Cognitive Demands: Why Revised Bloom's Taxonomy

Analyzing cognitive demands in a curriculum can map out what children are expected to learn and be able to do throughout their period of formal schooling. The question now is which tools are available or effective to analyze the dimensions and types of knowledge/skills required to teach/learn in schools. There are several tools available to analyze learning objectives in curriculum such as Bloom's Taxonomy (Bloom et al. 1956) and its revised version (Anderson et al. 2001), Klopfer's model (1971), the Structure of the Observed Learning Outcome (SOLO) taxonomy (Biggs 1995) and the New Taxonomy (Marzano and Kendall 2007). It is clear that each tool has its unique purposes, strengths, and limitations, and scholars using these tools have explained how these could help to understand cognitive dimensions of curriculum demands, students' learning process, and learning outcomes by compensating and overcoming the limitations of other tools. After scrutiny of these tools, revised Bloom's Taxonomy (RBT) was chosen as the most appropriate here, although it is not perfect for analyzing and comparing the cognitive dimensions of curriculum among the six states. In this section, we give an overview of those tools to explain why we chose RBT as the most suitable one for our study.

Bloom's Taxonomy (Bloom et al. 1956) was a widely used tool to characterize instructional objectives in educational documents such as curricula, perform objectives-based evaluation on students' achievement, and for aligning curriculum and assessment. The structural dimensions of knowledge and cognitive skills provided educators and policy developers practical ways to examine critical information on pedagogy and school practice for students' learning and cognitive development (Wee et al. 2011). Despite its practicality and accessibility, the early version of Bloom's Taxonomy was criticized for generalized and unidimensional domains of knowledge and skills that could not clearly explain the levels and dimensions of cognitive demands in the analyses of instructional objectives for

© The Author(s) 2017
Y.-J. Lee et al., *East-Asian Primary Science Curricula*,
SpringerBriefs in Education, DOI 10.1007/978-981-10-2690-4_2

students' learning and assessment plans. Furthermore, this version was also criticized for not being able to address the emphasis of higher levels of thinking, which became the focus of science teaching in the 1980s. These constraints led other tools to emerge such as Klopfer's model and the updated version of Bloom's Taxonomy.

Klopfer's model (1971) was designed to overcome criticisms concerning the generalized format of the original version of Bloom's Taxonomy (Bloom et al. 1956). It suggested domain-specific cognitive dimensions of science such as science inquiry and attitudes of science to emphasize students' learning beyond conceptual knowledge. A couple of researchers have employed this model as an alternative to analyze science curriculum. However, they found that the knowledge dimension in this model was also too broadly categorized in a one-dimensional structure, and it was separated from inquiry processes (Wee et al. 2011). This tool was likely to lead curriculum analysis into two separate domains, thus, unable to unpack the complexity of knowledge and cognitive processes. Anderson and Krathwohl (part of the original Taxonomy authors) recognized the need for revising the original version to overcome the limitations of one-dimensional categorizations, generalizations of cognitive dimensions, and difficulties of coding cognitive levels based on rigid hierarchical categorizations (Anderson et al. 2001; Krathwohl 2002). They revised Bloom's Taxonomy and published their work, *Revised Bloom's Taxonomy* in 2001. Compared to the previous version published 60 years ago, RBT has in essence made 12 changes (four each in emphasis, terminology and structure) adding another dimension of knowledge and interchanging the ranks of the highest two cognitive processes (i.e., Evaluate and Create). The revision accordingly proposed a two-dimensional approach to map cognitive development (i.e., adding knowledge dimensions and cognitive process) and also suggested verb forms (Understand, Explain, Infer, Create, etc.) to describe the cognitive processes with greater clarity. Thus, we now have four levels of knowledge: Factual, Conceptual, Procedural, and Metacognitive as well as six categories of cognitive processes, namely, Remember, Understand, Apply, Analyze, Evaluate, and Create.[1] Since then, RBT has been used internationally in mathematics and science curriculum research (e.g., Ari 2011; Porter 2006), although this literature has been extremely limited with regards to work in elementary science.

While RBT was under development, Marzano (2000) and his colleagues (Marzano and Kendall 2007) recognized the limitations of Bloom's Taxonomy and strove to develop a better tool for analyzing cognitive demands that they called the "New Taxonomy." They attempted to solve the issues of higher level thinking and sophisticated knowledge dimensions by looking into students' cognitive perception toward knowing and learning process. They emphasized the level of self and metacognitive domains of learning. This model consists of three systems (i.e., self-system, metacognitive system and cognitive system) and knowledge domains.

---

[1]These four dimensions of knowledge and six cognitive processes will be in capitals throughout this book. Thus, for example, we will simply state that an item is in Conceptual without attaching the word "category" after it.

They explained that when students are engaged in a new task or unfamiliar ideas, they always go through the initial process of cognition, that is, students perceive and pay attention to the task and ideas. This step is called self-system. Then later, students pay attention to what they would like to learn and how they achieve their goals, which is a dimension of the metacognitive system. And students process necessary information (cognitive system) to learn and accomplish their task. In this process, knowledge provides the content of thinking. This cognitive system is broken down to four components: Knowledge retrieval, comprehension, analysis, and knowledge utilization (Marzano 2000). Similar to RBT, the New Taxonomy also suggested a two-dimensional model to understand the complexity of students' learning and thinking. This model suggests an important lens to look into how students behave encountering new tasks, and it is critical to recognize the importance of self and metacognitive systems in students' learning. Yet self- and metacognitive systems in this model focus on the beginning part of students' learning process, whereas instructional objectives in curriculum focus on the targeted goals and outcomes of learning. Since our study of curriculum analysis does not have access to understanding students' learning behaviors or the process of cognitive development over time, the New Taxonomy is not the best tool to deliver our research goals.

The SOLO taxonomy is another popular tool to comprehend cognitive development in students' learning. This model proposes several dimensions of cognition; prestructural, unistructural, multistructural, relational, and extended abstract to understand the complexity and nonlinearity of children's cognitive development. It is particularly useful for evaluating the growth of students' cognition in classrooms (Biggs 1995). That is, it assesses what levels of cognition students develop over time. For instance, teachers can evaluate whether a learner can list the stages of life cycles of a butterfly (unistructural) before being able to explain the complexity and interrelation of life cycles and habitats (relational). Thus, this tool is more appropriate to assess process of children's cognitive development in classrooms (Brabrand and Dahl 2009) rather than analyzing intended cognitive demands in curriculum.

Given that this study intends to analyze and compare the learning objectives stated in curriculum documents, the RBT was chosen as the most appropriate tool. SOLO and the New Taxonomy are better for examining students' behaviors and learning progression while Klopfer's model was not able to unpack the complexity of students' learning. There are also some tools developed in local regions such as semantic network analysis in a study of Korean curriculum analysis (Chung et al. 2013) that suggested the importance of being critical toward various tools to analyze learning objectives in curriculum. Yet due to the regional uniqueness and language, those tools also had their own limitations. Understanding the strengths and weakness of each tool, we decided to employ revised Bloom's Taxonomy (RBT) as the most suitable and effective tool to reach the goals of our study.

We do not claim that RBT is a perfect tool to analyze cognitive dimensions of learning objectives. We acknowledge there are certain limitations in RBT. As students' thinking and learning is a complex, nonlinear notion, the hierarchical approach in RBT regarding the levels and dimensions of knowledge and process might be too simplistic to explain the complexities of cognition. In this regard, it is important for us to further unpack some of the problems with RBT and use the tool with mindfulness and clarity.

Further Critiques of Bloom's Taxonomy

In a chapter on reading comprehension research in science, Holliday and Cain (2012) report that popular strategies in this domain have been largely ineffective, including those that all rely on Bloom's Taxonomy:

> Bloom's taxonomy sounds like a good idea, except for the lack of supportive evidence-based theory and research, as noted by Richard C. Anderson (1972). More generally, it is amazing how long myths can stay alive, resulting in preservice teachers in universities being taught and tested about, for example, Bloom's six dreamed-up levels and their mythical hierarchy (p. 1406).

They concede, however, that "if these…notions seem to work for you, then perhaps you should not categorically dismiss them" (p. 1406). Such strong criticisms of Bloom's Taxonomy, in fact, are not new and have been around for a number of decades. Most of them revolve around the empirical basis for the hierarchy of processes, the categories, and questions about its efficacy in/for curriculum work (see Anderson and Sosniak 1994). Even after the publication of the revised version of the Taxonomy, such doubts have continued although there is now room for some optimism. For example:

> The results of the meta-analysis provide slight support for the cumulative hierarchy assumption….Our re-analysis supports the conclusions reached in most of the individual studies–namely, that, excluding knowledge, the support for a cumulative hierarchy is clearest for the simpler categories of the structure. Kreitzer and Madaus (1994) reached this same conclusion (Anderson et al. 2001, p. 293).

Other problems regarding its rather theoretical or abstract manner for curriculum-making have been perhaps overstated; in particular, there have been a number of concerns regarding its application for curriculum development, instruction as well as for assessment (see Anderson et al. 2001, Chap. 17). Looking back over the 60 years since the first version of the Taxonomy was published, it is nonetheless insightful to read how Benjamin Bloom himself cogently justified the Taxonomy for educational purposes.

> The Taxonomy does not impose a set of teaching procedures, nor does it view objectives as so detailed and restrictive that a single teaching method is implied. Rather, a teacher has a wide range of choices in making instructional decisions related to objectives associated with each level of the Taxonomy. The Taxonomy does emphasize the need for teachers to help students learn to apply their knowledge to problems arising in their own experiences and to be able to deal effectively with problems that are not familiar to them. This emphasis

alone should guard against the rote learning of ready-made solutions. It is obvious, at least to me, that many of the criticisms directed towards the Taxonomy have resulted from very narrow interpretations of both the Taxonomy and its proper applications (Bloom 1994, p. 7).

As he indicated, some problems could take place in its narrow interpretations and implications in educational settings, but it is also critical to recognize the limitations of the hierarchical categorization of cognition in educational practice such as enactive learning and teaching. Despite the two-dimensional structure of revised Bloom's Taxonomy attempted to overcome the generalization of cognitive dimensions, there still remains the challenge of analyzing the level of thinking (Marzano and Kendall 2007). For instance, Evaluate is a higher level than Analyze in the hierarchy of RBT, but the latter often requires more complex cognitive process than a simple evaluate activity. That is, a learning objective such as "students draw a conclusion from data of population changes and environmental changes" (Analyze) requires a higher and more complex level of cognition than "students can judge if temperature affects plant growth" (Evaluate) (Anderson et al. 2001). Thus, just looking into the verbs of learning objectives in the hierarchy system is sometimes limited when explaining the level of cognitive demands. Thus, coding always requires careful interpretations through the Taxonomy. Furthermore, RBT aims to identify only knowledge and skills. That is, it is not able to understand students' learning process, values, and attitudes in students' learning. These latter categories are also critical aspects of science literacy and goals of science education, but there are no categories to recognize these in RBT. In this regard, RBT does not capture the holistic view of instructional objectives within science curricula.

Despite all these limitations and problems, RBT is still an extremely useful tool when used for specific purposes such as understanding cognitive demands of instructional objectives in curriculum documents, which is the intention of our study. As one of the few educational innovations that have come from academia and found an enormously welcoming acceptance in K-12 systems in the USA and even more overseas, Bloom's Taxonomy is a well-loved and respected tool despite known shortcomings (Schneider 2014). More so when the revised version has been intentionally updated for extensive adoption within K-12 school systems, a departure from its original domicile of higher education. We like to think that in some respects, RBT is analogous to how scientists have viewed the structure of the atom; there are elaborate models/theories backed by various degrees of empirical evidence, but the exact structure of an atom can never be known with complete certainty. It is perhaps too cynical to label all these categories in RBT as mere hypothetical constructs or useful fictions as some are wont to do given the immense contribution that the Taxonomy has functioned as a curriculum tool over the past 60 years. As long as researchers recognize the limitations of RBT and are clear about their research intentions with this tool, using RBT can be an effective method to serve their purpose.

References

Anderson, L. W., & Sosniak, L. A. (Eds.). (1994). *Bloom's taxonomy: A forty-year retrospective. Ninety-third yearbook of the National Society for the Study of Education. Part II.* Chicago, Il: The University of Chicago Press.

Anderson, L. W., Krathwohl, D. R., Airasian, P. W., Cruikshank, K. A., Mayer, R. E., Pintrich... Wittrock, M. C. (Eds.). (2001). *A taxonomy for learning, teaching and assessing. A revision of Bloom's taxonomy of educational objectives.* White Plains, NY: Addison-Wesley Longman.

Ari, A. (2011). Finding acceptance of Bloom's revised cognitive taxonomy on the international stage and in Turkey. *Educational Sciences: Theory and Practice, 11*, 767–772.

Biggs, J. (1995). Assessing for learning: Some dimensions underlying new approaches to educational assessment. *The Alberta Journal of Educational Research, 41*, 1–17.

Bloom, B. S. (1994). Reflections of the development and use of the Taxonomy. In L. W. Anderson, & L. S. Sosniak (Eds.), *Bloom's taxonomy: A forty-year retrospective. Ninety-third yearbook of the National Society for the Study of Education. Part II* (pp. 1–8). Chicago, Il: The University of Chicago Press.

Bloom, B. S., Engelhart, M. D., Furst, E.J., Hill, W.H., & Krathwohl, D., R. (1956). *Taxonomy of educational objectives: The classification of educational goals. Handbook I: Cognitive domain.* New York: David McKay.

Brabrand, C., & Dahl, B. (2009). Using the SOLO taxonomy to analyze competence progression of university science curricula. *Higher Education, 58*, 531–549.

Chung, D. H., Lee, J.-K., Kim, S. E., & Park, K. J. (2013). An analysis on congruency between educational objectives of curriculum and learning objectives of textbooks using semantic network analysis-focus on Earth Science I in the 2009 revised curriculum. *Journal of the Korean Earth Science Society, 34*, 711–726.

Holliday, W. G., & Cain, S. D. (2012). Teaching science reading comprehension: A realistic, research-based approach. In B. Fraser, K. Tobin, & C. McRobbie (Eds.), *Second international handbook of science education* (pp. 1405–1417). Dordrecht: Springer.

Klopfer, L. E. (1971). Evaluation of learning in science. In B. S. Bloom, J. T. Hastings, & G. F. Madsus (Eds.), *Handbook on formative and summative evaluation of students learning.* NY: McGraw-Hill.

Krathwohl, D. R. (2002). A revision of Bloom's taxonomy: An overview. *Theory into Practice, 41*, 212–264.

Marzano, R. J. (2000). *Designing a new taxonomy of educational objectives.* Thousand Oaks, CA: Corwin Press.

Marzano, R. J., & Kendall, J. S. (Eds.). (2007). *The new taxonomy of educational objectives.* Thousand Oaks, CA: Corwin Press.

Porter, A. C. (2006). Curriculum assessment. In J. L. Green, G. Camilli, & P. B. Elmore (Eds.), *Handbook of complementary methods in education research* (pp. 141–159). Mahwah, NJ: Lawrence Erlbaum.

Schneider, J. (2014). *From the ivory tower to the schoolhouse: How scholarship becomes common knowledge in education.* Cambridge, MA: Harvard Education Press.

Wee, S.-M., Kim, B.-K., Cho, H., Sohn, J., & Oh, C. (2011). Comparison of instructional objectives of the 2007 revised elementary science curriculum with 7th elementary curriculum based on Bloom's revised taxonomy. *Journal of Korean Elementary Science Education, 30*, 10–21.

Chapter 3
Curricula, Coders, and Coding

Sources of Curricula Material

The study began with collecting official curriculum documents from each state. Bearing in mind that we referenced only the learning objectives or standards in the cognitive domain while ignoring those in the affective domain, the details of the sources of these curricula are as follows:

- The primary science learning objectives (which are part of the subject known as Integrated Studies) in Hong Kong is available online (http://www.edb.gov.hk/attachment/en/curriculum-development/kla/sen/ScKLA-e.pdf) so we downloaded the document and confirmed with a Hong Kong scholar that the 2002 version was the most current science curriculum being used in Hong Kong.
- The current Japanese curriculum was released in March 2008 and fully implemented in April 2011. The curriculum analyzed in this study is the English (tentative) version of this current curriculum 2011. The English translation was developed by the ministry and is available on the ministry's official website. (http://www.mext.go.jp/component/a_menu/education/micro_detail/__icsFiles/afieldfile/2009/04/21/1261037_5.pdf). The original Japanese version is found in https://www.nier.go.jp/guideline/h19e/chap2-4.htm
- The Chinese curriculum from mainland China was released in 2001 and the document is available at the website of People's Education Press (http://www.pep.com.cn/xxkx/xxkxjs/kbjd/xkkcbz/201009/t20100901_851903.htm. There is a newer version of curriculum document (2013) but it is still under examination and revision in the country and not officially released yet. Thus, we analyzed the 2001 science curriculum.
- The Taiwanese curriculum was obtained from the website of the Taiwanese Ministry of Education (http://www.k12ea.gov.tw/ap/sid17_law.aspx). It was published in 2008 with some minor corrections, and a new version is expected to be released probably in 2018.

© The Author(s) 2017
Y.-J. Lee et al., *East-Asian Primary Science Curricula*,
SpringerBriefs in Education, DOI 10.1007/978-981-10-2690-4_3

- Korean curriculum documents that we analyzed in this study are the most recent among the six states. The last science curriculum document was in 2009. During the reforms from 2014 to 2015, the new 2015 version was released in September, 2015. One of the researchers (HGY) was one of the curriculum developers and could provide the printed document as well as critical knowledge of curriculum reforms in Korea. See http://ncic.go.kr/
- The Singapore curriculum was a publicly available document and thus obtained online (http://www.moe.gov.sg/education/syllabuses/sciences/files/science-primary-2014.pdf). This is the latest version of the primary science curriculum (2014), the last revision was in 2008. All cognitive and skills-based objectives for pupils studying in the standard (not foundation) stream were chosen, which constitute the majority of learners in the country.

Selection of Learning Objectives or Standards

When we collected official/national curriculum material from each state, we selected learning objectives that only dealt with cognitive processes (including items on process/inquiry skills) and ignored those that dealt with values and attitudes. As mentioned, we selected only those objectives that related to what typically should belong to a primary science curriculum for the case of Hong Kong, which other scholars might dispute our selection from the General Studies subject. All our coding and selection choices are open to inspection upon request; we have not published the official learning objectives in this book (other than from Singapore) due to copyright restrictions.

There are numerous fine differences in how curriculum standards, learning objectives/competencies, and the like are defined around the world, even among states or districts within a country. For instances, some states use the term achievement standards (e.g., Korea) for learning objectives, or learning outcomes (e.g., Singapore), and yet others learning standards (e.g., mainland China). These variations of meaning—explicit or implicit—make comparative research like ours easily susceptible to misinterpretation by both producers and consumers of such research (see Waddington et al. 2007). For instance, while a set of content standards might not be prescriptive in terms of the choice of pedagogical methods, teachers might still regard the specified content to index the taught curriculum, that is, levels of learning attainment now dictate what is to be covered. In this study, we are confident that this is indeed the case across all six states regardless if content standards were the phrase of choice in curriculum documents found in China and Taiwan (specifically, "guidelines" was used in these places). We make this claim given the high degree of specificity of learning attainment and the strong presence of testing in all six states together with the high adherence to teaching the entire intended curriculum by local teachers.

The Selection of Coders

As for coders of curriculum, we chose coders based on their mother/native tongue first and later we took into consideration their second language and convenience of face-to-face meetings. For example, for the Korean curriculum, there was no English version of the curriculum, thus two researchers whose first languages is Korean coded the curriculum. For the Chinese curriculum, Qingna Jin's first language is Chinese and she is in Canada working with Mijung Kim. Thus, even though Mijung Kim does not speak Chinese, Qingna translated the curriculum into English and they met several times to discuss the coding strategies and difficulties of translation. Thus it was very convenient to develop coding together. For the Taiwanese curriculum, Yew-Jin and Qingna coded together since Yew-Jin's second language is Chinese. We also attempted to refine all our English translations of the learning objectives from the vernacular several times keeping in mind the original intent of the objectives while also phrasing it consistently and in terms or ways that international scholars would recognize. Thus, our new English translations of the intended primary science learning objectives (cognitive domain) are provided in the Appendices for China, Taiwan, and Korea, which we believe will be a very useful contribution for scholars.

The Coding Process and Interrater Reliabilities

Because this was a comparative study of learning objectives using an instrument that can be subject to numerous judgment calls as well as the involvement of five coders spread over four states, we had to undergo a lengthy process of training in the coding procedures in RBT. The process of coding was based on previous research by Lee et al. (2015). Pairs of researchers (that included at least one native speaker of the language) were tasked to analyze the learning objectives from each of the six East-Asian states according to RBT: Hong Kong (YJL & MK), Japan (YJL & KM), China (MK & QJ), Taiwan (YJL & QJ), Korea (HGY & MK), and Singapore (YJL & HGY). All objectives written in vernacular languages (e.g., from China, Taiwan) were also translated into English. The pairs independently practiced with a random sample of objectives ($\sim$ 10–16 items) from another East-Asian state before jointly engaging in discussion to understand their differences and resolve peculiarities in coding.

Their inter-reliability scores–Cohen's kappa (https://en.wikipedia.org/wiki/Cohen%27s_kappa) and percentage agreements–at this stage are thus shown in Tables 3.1 and 3.2 (Graham et al. 2012). Kappa takes into account the agreement or disagreement by chance between two coders that adds extra value compared to just calculating percentage agreement in a study. It is also enhanced when there are more coding categories like in our case instead of yes/no for example. The same pairs of coders then proceeded to code their target objectives separately before reconciling any differences and achieving a final consensus for that state. For instance, JQ and YJL randomly picked 15 learning objectives from Korean curriculum and compared their coding results and discussed their own coding strategies. This allowed them to share the coding schemes and strategies in order to get ready to code the Taiwanese curriculum together. This process was repeated among all the researchers to result in the stable and consistent coding results across the countries.

Tables 3.1 and 3.2 show that the Cohen's kappa values and percentage agreement levels varied among countries and this could be explained with several reasons. The coding of the Korean curriculum was done after we had finished coding the five other curricula. This means that the coders of the Korean curriculum, HGY, and MK had been involved in online conversations and discussions for several months while the other researchers were coding their curricula. Whenever there were questions and disagreement on coding, they discussed intensively through online conversations (emails, skypes, and google document writing) to resolve misunderstandings, conflicts, and disagreement on coding schemes. Based on these conversations, by the time HGY and MK started their coding in Sept 2015, they were very clear and consistent about the coding schemes; thus, the percentage agreement rates were high (though not kappa for knowledge).

In the case of the Chinese curriculum, QJ and MK met several times to discuss the dimensions of cognitive demands in Bloom's Taxonomy because it was the first time for Qingna Jin to learn about the tool. QJ and MK then randomly picked and coded learning objectives to learn how RBT could be utilized in actual coding. Also, they coded learning objectives in the draft 2013 Chinese curriculum first before they started coding the official 2001 curriculum and thus it was not surprising that the percentage agreement rate was high. In the case of Japan, kappa for the cognitive domain was 0 although percentage agreement was high due to how the formula of kappa was calculated; there were many empty cells that affected its final value. This unusual phenomena have been reported by many other researchers although Cohen's kappa is still very popular in the social sciences (http://www.eadph.org/congresses/16th/Mimimum_reporting_requirements.pdf). Nonetheless, with few learning objectives ($n = 31$) any small number of disagreements could severely affect the interrater reliabilities, which seemed to be the case in the coding

Table 3.1 Table showing the interrater values of kappa and percentage agreement for the knowledge domain

States	Interrater reliabilities for Knowledge	
	Cohen's kappa (SE in brackets)	Percentage agreement (%)
Hong Kong	0.67 (0.09)	78.9
Japan	0.18 (0.16)	45.9
China	0.74 (0.04)	82.8
Taiwan	0.87 (0.04)	92.1
Korea	0.46 (0.12)	85.0
Singapore	0.41 (0.08)	61.4

Table 3.2 Table showing the interrater values of kappa and percentage agreement for the cognitive domain

States	Interrater reliabilities for cognitive processes	
	Cohen's kappa (SE in brackets)	Percentage agreement (%)
Hong Kong	0.39 (0.08)	54.3
Japan	0.00 (0.55)	90.3
China	0.63 (0.04)	74.9
Taiwan	0.74 (0.05)	81.6
Korea	0.89 (0.04)	95.0
Singapore	0.44 (0.08)	65.1

of the Japanese curriculum for the knowledge dimension. To some extent, we believe this happened for the cognitive domain for Hong Kong too. Save for Japan, kappa values for knowledge ranged from 0.41 (moderate) to 0.87 (almost perfect agreement) and 0.39 (fair agreement) to 0.89 (almost perfect agreement) for the cognitive processes, which we felt was satisfactory for this conservative agreement statistic. It is to be further noted that even with more experienced coders, one does not necessarily get high kappa interrater reliability values. The Korean coders who coded in the original languages and who had much prior experience in coding had lower kappa values for knowledge than the coders for the curriculum in China where JQ joined the study much later. Again, this is another instance of the peculiarity of kappa calculations.

Advice and Tips for Coding Using RBT

Apart from defining in detail the four types of knowledge and six categories of cognitive processes, Anderson et al. (2001) were helpful in providing practical tips and advice for research coding using RBT, which we briefly summarize below with reference to our study:

(i) Paying attention to verb (i.e., process) and noun (i.e., knowledge) phrases

We tried to be as consistent as possible by identifying and understanding the verbs and nouns from the objectives in a similar and rather literal way. This was partly achieved when the pairs of coders practised coding with items from our bounded pool of six states as an initial step before coding those from the target state on our own. We quickly came to realize that verbs such as "Know," "Recognise," "List," or "State" were best classified as Remember unless there was reason to suggest otherwise or when these were combined with a more demanding command verb within the same objective. The verb "Know" was especially troublesome in Chinese and Korean curriculum documents as it was signified by a few closely allied verbs (知道, 认识, 了解 in Chinese and 안다 in Korean) and thus needed to be compared consistently with similar items in the curriculum. We too were sensitive to verbs such as "Distinguish" or "Differentiate" that were most likely to be associated with Analyze.

As in the previous study (Lee et al. 2015), whenever two verbs were present we coded it according to the more demanding cognitive process. Hence, if asked to "identify the organs in the human digestive system [mouth, gullet, stomach, small intestine and large intestine] and describe their functions," this was regarded as asking learners to understand (i.e., to describe rather than merely recall) conceptual information, here the functions of organs in the digestive system.

Mapping the subprocesses and exemplar command verbs from Anderson et al. (2001) with the objectives proved very useful whenever we were debating over ill-defined or fuzzy objectives that had subtle meanings in the original language. Other important contextual clues such as the expected grade level of the objective, word length of objective, and science topic or activity were also instrumental in our decision-making.

During the coding for knowledge, some degree of inference was inevitable to pinpoint what types of learning outcomes were intended as we could not bring complementary insights from the local-enacted and tested curriculum for some states. Still, disputes about knowledge levels were generally easier to harmonize as there were just four groups compared to six processes of cognition.

The coding of procedural knowledge was surprisingly troublesome; we deliberated if we should code for the ultimate/desired learning outcomes (in terms of knowledge levels) or the knowledge levels that were achieved/required for the very process of that learning. This interesting conundrum was already known to the authors of RBT who stated that very often the result of using procedural knowledge was actually factual or conceptual knowledge. Keeping in mind that the pairing of Apply-Procedural was supposed to be commonplace, we can illustrate our process of decision-making here by the following pair of hypothetical learning objectives:

- Students use microscopes to learn how to properly observe small, invisible objects such as cells [LO1]
- Students use microscopes to learn about cell structures in plants, e.g., the nucleus, cell wall, cytoplasm [LO2].

In both objectives, the verb "use" is typically tied with Apply and is not in serious dispute. However, in objective 1 [LO1] the noun phrase "learn how to properly observe small, invisible objects such as cells" we are concerned with mastering practical knowledge of a specific method or technique called microscopy–this falls squarely under the ambit of procedural knowledge. On the other hand, in objective 2 [LO2] the noun phrase "learn about cell structures in plants" implies knowledge of which cellular structures are found in plants and their characteristics, which are linked with conceptual knowledge. In the former objective [LO1], assessment would, for example, test if students can use the microscope without error while in the latter teachers would check what students know about plant cells. We have decided that although procedural knowledge was definitely required in the second objective, we wanted to be faithful to the way it was phrased as a written intended learning objective. Objectives that required learners to know (e.g., mental division) or do something (e.g., show calculations on paper) with subject-specific skills, algorithms, methods, techniques, or procedures thus belonged to Procedural—none of which were described in objective 2 [LO2]. In conclusion, identifying and understanding the correct verb–noun combinations were the one of the most fundamental operation in our method of coding.

(ii) Relating types of knowledge to processes

It was imperative for the coders to pay attention that there were certain knowledge-process combinations that were more prevalent in educational contexts, that is, we are very likely to observe more Remember-Factual, Understand-Conceptual, and Apply-Procedural pairings. This general correspondence that was observed since the first version of RBT is not so well obeyed with the more demanding processes such as Analyze, Evaluate, and Create that could be readily linked with any of the four Knowledge groups. Moreover, it was expected that coding for "Comprehension," which has now been renamed as Understand was "probably the largest general class of intellectual abilities and skills emphasized in schools and colleges" (Bloom et al. 1956, p. 89).

(iii) Teaching activities need to be differentiated from learning objectives

This principle was very problematic for the research team as a whole because all of us were classroom instructors at some point and thus we tended to code the intended learning objectives as how we would actually teach them! Also, some of us were actually involved in textbook development in the country and most of us are familiar with science textbooks and teacher guidebooks which are used as part of curriculum or as actual curriculum in classrooms; thus, it was difficult to separate what teachers and students actually do from the learning objectives of the curriculum. This normative pedagogical stance as long-time practitioners was difficult to erase and we had to consciously remind ourselves to code what we saw in the curriculum documents rather than how we would teach the former—this was after all, the intended curriculum. An exception was made for a set of 26 objectives from Singapore that were officially listed in the section of Skill and Processes (see CPDD 2013). The coders reached a compromise in that the minimal knowledge level for coding these items was agreed to be at the level of Procedural, whereas these objectives could be freely linked with any type of cognitive process. In general, the insidious influence of a teacher's past experience of or intent for instruction should not be underestimated during the coding of curriculum objectives as has been our experience.

(iv) Specificity of objectives and prior learning

Anderson et al. (2001) further described how the level of specificity of learning objectives and ascertaining the prior learning of learners was crucial. We were helped with respect to the former concern in that these learning objectives were already rather precisely written by governments as these were meant to be implemented at the national level, which required clarity as well as accuracy. While we were unable to ascertain prior learning of students, this aspect of coding was inconsequential as we were analyzing objectives at the national level, not individual classrooms.

References

Anderson, L. W., Krathwohl, D. R., Airasian, P. W., Cruikshank, K. A., Mayer, R. E., Pintrich... Wittrock, M. C. (Eds.). (2001). *A taxonomy for learning, teaching and assessing. A revision of Bloom's taxonomy of educational objectives*. White Plains, NY: Addison-Wesley Longman.

Bloom, B. S., Engelhart, M. D., Furst, E.J., Hill, W.H., & Krathwohl, D., R. (1956). *Taxonomy of educational objectives: The classification of educational goals. Handbook I: Cognitive domain*. New York: David McKay.

CPDD [Curriculum Planning & Development Division]. (2013). *Science syllabus primary 2014*. Singapore: CPDD: Ministry of Education. Retrieved from https://www.moe.gov.sg/docs/defaultsource/document/education/syllabuses/sciences/files/science-primary-2014.pdf

Graham, M., Milanowski, A., & Miller, J. (2012). *Measuring and promoting inter-rater agreement of teacher and principal performance ratings*. Madison, WI: Center for Educator Compensation Reform, University of Wisconsin-Madison.

Lee, Y.-J., Kim, M., & Yoon, H.-G. (2015). The intellectual demands of the intended primary science curriculum in Korea and Singapore: An analysis based on revised Bloom's taxonomy. *International Journal of Science Education, 37*, 2193–2213.

Waddington, D., Nentwig, P., & Schanze, S. (Eds.). (2007). *Making it comparable: Standards in science education*. Munster, Germany: Waxmann.

Chapter 4
The Intellectual Demands of East-Asian Primary Science Curricula

Overall Cognitive Processes and Knowledge Levels Across Six States

Tables 4.1 and 4.2 show our coding of the cognitive processes and levels of knowledge of the intended primary science curricula across all the six East-Asian states respectively. It is to be remembered that some degree of subjectivity is inevitable when coding the intellectual demands, especially with the large number of learning objectives or standards that we selected belonging to the cognitive domain. In total, these amounted to 560 objectives/standards from the six states.

In the cognitive process of Remember, three states (China, Hong Kong, Singapore) garnered high amounts of representation here with about a third of their curriculum (cognitive domain only) inside this category. As to be expected, most of the East-Asian learning objectives were very well represented in Understand; most states had at least a quarter to a third of their objectives located here with the singular exception of Japan. In fact, in the Japanese curriculum nearly every objective was located within Apply, which paralleled the strong national emphasis on applied knowledge of science during classroom teaching and learning. Seen from another angle, this emphasis on hands-on activities in Japan led to the absence of items in Remember, Understand, Evaluate, and Create. Korea too recently had a strong emphasis on learning science through hands-on activities; Apply therefore had 62 % of all objectives located here. Items in Analyze, Evaluate, and Create in the RBT are considered to require higher order thinking skills; it was observed that most states had modest frequencies of items (~ 5 %) in these three categories. Of note, Taiwan had 10 % of their learning objectives in Create, which was the highest figure among the six states followed closely by Hong Kong with 7 %. Only three states—Korea, Japan, Singapore—lacked at least one or more objectives within any of the six cognitive processes in RBT.

© The Author(s) 2017
Y.-J. Lee et al., *East-Asian Primary Science Curricula,*
SpringerBriefs in Education, DOI 10.1007/978-981-10-2690-4_4

Table 4.1 The overall profile of learning objectives/standards in % from six East-Asian states classified by the cognitive processes in RBT

	Remember	Understand	Apply	Analyze	Evaluate	Create
Hong Kong (n = 57)	31.6	31.6	21.1	5.2	3.5	7.0
Japan (n = 31)	0	0	93.5	6.5	0	0
China (n = 162)	29.6	38.9	19.8	3.7	3.1	4.9
Taiwan (n = 114)	16.7	24.6	38.6	5.3	4.3	10.5
Korea (n = 113)	0	34.5	62.0	0	0	3.5
Singapore (n = 83)	31.4	32.5	31.4	4.8	0	0

Table 4.2 The overall profile of learning objectives/standards in % from six East-Asian states classified by the knowledge domains in RBT

	Factual	Conceptual	Procedural	Metacognitive
Hong Kong (n = 57)	15.8	49.1	33.3	1.8
Japan (n = 31)	0	74.2	25.8	0
China (n = 162)	26.5	54.3	18.6	0.6
Taiwan (n = 114)	16.7	35.1	47.4	0.9
Korea (n = 113)	0.9	79.6	19.5	0
Singapore (n = 83)	16.9	67.5	15.6	0

In terms of the knowledge dimension, Conceptual occupied the lion's share of objectives from every state (ranging from 35 to 79 %), a pattern that is to be expected (other than Taiwan). In particular, Korea, Japan, and Singapore were extremely well represented in this category. It is interesting to note that Factual items (other than from China) were not plentiful and ranged from about 16 % to nearly 0 % for Korea and Japan. Procedural knowledge too was well represented in the Taiwanese curriculum; items here occupied nearly half of all learning objectives from this state while the other states had between 20 and 30 % representation in this knowledge group. The category of Metacognitive was very poorly represented in all six states. Again, only three states—Korea, Japan, Singapore—lacked at least one objective within the four knowledge domains in coding of RBT.

It is interesting to note that almost two-thirds of the learning objectives belonged both to Apply and Procedural in the Korean curriculum. This result came from the unique structure of learning objectives, that is, many objectives adopted the form of by/through doing certain scientific process students will achieve certain types of knowledge. In this form, students are engaged in Apply processes and Procedural knowledge. In Japan, the recommended teaching method whereby students had to do something in order to learn scientific ideas was even more stark; 93.5 % of Japanese objectives were in Apply and 74.2 % in Conceptual! We now describe the findings of our codings for each of the six states in turn.

Hong Kong

General Introduction

Having a high-performing education system, Hong Kong presents many insightful lessons regarding curriculum reforms interacting with larger sociopolitical changes, not just within the discipline of science education (e.g., Kennedy 2005; Marsh and Lee 2014). Primary science (i.e., science and technology education) is part of the General Studies (GS) curriculum in Hong Kong SAR, which combines learning about science, health education, and social studies through inquiry. Published in 2002, GS was fully implemented in 2004 and consists of six strands, namely:

- Health and Living
- People and Environment
- Science and Technology in Everyday Life
- Community and Citizenship
- National Identity and Chinese Culture
- Global Understanding and the Information Era (see Curriculum Development Council [CDC] 2011).

Moreover, the aims of GS are the following:

- to provide learning experiences for students to have a better understanding of themselves and the world around them
- to arouse students' interest in and develop their skills to enquire about themes and issues related to science, technology, and society
- to cultivate positive attitudes and values for healthy personal and social development (CDC 2011, p. iii).

Organized into Key stages 1 (Grades 1–3) and 2 (Grades 4–6), GS occupies 12–15 % of total curriculum time or slightly over 3 hours per week to cover the six GS strands. Because this is an integrated curriculum, we took what we deemed as relevant learning objectives in science, technology, and human health from the first three strands listed above. The list of our choices deemed typical of primary science learning is available upon request. The overarching curriculum framework in Hong Kong consists of three elements: Eight Key Learning Areas (KLA, which comprises the typical subjects), nine generic skills, and various values and attitudes. It should be patently clear by now that curriculum in Hong Kong expresses and is overlaid with various aims, objectives, and goals of which not every one has been described in this chapter. This highly sophisticated conceptualization of what is an educated person ready for life in a globalized world has led Robin Alexander, noted comparative researcher of primary education systems to declare that:

> I have to say that this is one of the more complex of the hybrid specifications that I have seen, because although it starts as the familiar two-dimensional grid (key learning areas – or subjects – and generic skills) it actually has six dimensions. This degree of complexity raises the stakes when it comes to implementation, and makes it possible that some

elements, in some schools, will be delivered more as rhetoric than practice, for it is hard to pursue so many objectives simultaneously. (Alexander 2008, p. 149)

Hong Kong teachers are encouraged to contextualize the GS curriculum to address the needs of each school as well as use different approaches and strategies. It comes as no surprise that school-based curriculum development features very large here when teachers are told to devote 80 % of teaching time on the core curriculum, while the rest of the time they have the freedom to cater to specific interests and needs of their students. With respect to the teaching of science, attention should be paid to providing many hands-on learning experiences in and out of the classroom. This active learning approach was also to be coupled with the development of scientific thinking and critical reasoning in the subject (CDC 2002). In parallel to the GS curriculum, there are six strands in the science curriculum that are as follows:

- Scientific Investigation—to develop science process skills and understanding of the nature of science
- Life and Living—to develop understanding of scientific concepts and principles related to the living world
- The Material World—to develop understanding of scientific concepts and principles related to the material world
- Energy and Change—to develop understanding of scientific concepts and principles related to energy and change
- The Earth and Beyond—to develop understanding of scientific concepts and principles related to the Earth, Space, and the Universe
- Science, Technology and Society (STS)—to develop understanding of the interconnections between science, technology, and society (CDC 2002, p. 20).

Overall State Profile of Learning Objectives

Items in Conceptual alone accounted for half of all learning objectives in Hong Kong (see Table 4.3). The next most frequent items were in the Procedural domain which occupied a third of all items. With respect to processes, both Remember and Understand had 63.2 % of all items that was close to the number in the three pairings of Remember-Factual, Understand-Conceptual, and Apply-Procedural (68.4 %). Although the pair Conceptual:Understand was supposed to be the most ubiquitous, it just garnered 31.6 % of all items in Hong Kong.

Table 4.3 Table showing total number of learning objectives ($n = 57$) from Hong Kong (SAR) classified according to the dimensions of knowledge and cognitive processes in RBT

	Remember	Understand	Apply	Analyze	Evaluate	Create	Number of knowledge items
Factual	9 (100) (50.0)	0	0	0	0	0	9 (15.8)
Conceptual	8 (29.6) (44.4)	18 (66.7) (100)	0	1 (3.7) (33.3)	0	0	27 (49.1)
Procedural	1 (5.0) (5.6)	0	12 (60.0) (100)	2 (10.0) (66.7)	1 (5.0) (50.0)	4 (20.0) (100)	20 (33.3)
Metacognitive	0	0	0	0	1 (100) (50.0)	0	1 (1.8)
Number of cognitive items	18 (31.6)	18 (31.6)	12 (21.1)	3 (5.2)	2 (3.5)	4 (7.0)	57

Percentages shown in brackets (%)

Profile of Upper and Lower Primary Learning Objectives

It is interesting to note that as children in Hong Kong move up to Key Stage 2, the number of items in Conceptual increased nearly threefold at the expense of items in Procedural, and especially in Factual (Table 4.4). This might reflect a move away from teaching using hands-on learning activities toward learning of abstract conceptual knowledge although there is no way to ascertain this speculation from the learning objectives. For good reasons many policymakers wish not to place too great a burden on younger children on their initial encounters with science. As well, for curricula that adopt a spiral approach, simply counting the changes in the learning objectives might not provide the full picture of the intellectual demands that come with monitoring these numbers across grade levels; not all objectives are indeed the same.

The two Key Stages in Hong Kong are rather similar in terms of their cognitive profiles though again we can observe an increase in Understand, which is expected because of the large increase in Conceptual knowledge demands among items (see Table 4.5).

Table 4.4 Number of learning objectives in the knowledge domain from Hong Kong (SAR) sorted according to their grade levels

Key stage	Factual	Conceptual	Procedural	Metacognitive	Total ($n = 57$)
1.	8 (32.0)	6 (24.0)	11 (44.0)	0	25 (43.8)
2.	1 (3.1)	21 (65.7)	9 (28.1)	1 (3.1)	32 (56.2)

Percentages shown in brackets (%)

Table 4.5 Number of learning objectives in the cognitive domain from Hong Kong (SAR) sorted according to their grade levels

Key stage	Remember	Understand	Apply	Analyze	Evaluate	Create	Total (n = 57)
1.	9 (36.0)	6 (24.0)	6 (24.0)	1 (4.0)	1 (4.0)	2 (8.0)	25 (43.8)
2.	9 (28.1)	12 (37.5)	6 (18.8)	2 (6.3)	1 (3.0)	2 (6.3)	32 (56.2)

Percentages shown in brackets (%)

Japan

General Introduction

Education in Japan follows a 6-3-3 pattern starting with six years of primary education and three for lower secondary education thus making a total of nine years of compulsory education. Science lessons begin in the third grade and science is a required subject throughout compulsory education. In elementary schools, science is taught to all students in mixed-ability classes and pupils are not streamed/tracked. Unlike the situation for entry to high schools and universities, there is no entrance examination for public junior high schools.

The Japanese national curriculum—the Courses of Study—has been revised eight times since its implementation in 1947 to keep up with changes in society over the years and the learning needs of each age group. The current Course of Study for Elementary Schools, which includes the primary science curriculum, was announced in March 2008, and fully implemented from April 2011. The subject-specific sections of the Course of Study for Elementary Schools consist of three parts: Overall objectives; Objectives, and contents for each grade or section (objectives, contents and teaching (handling) of the contents), and Syllabus design and guidance for teaching (handling) the contents. In section 4 of the Course of Study for science, it lists the overall objectives of primary science, the objectives and contents for each grade (Grades 3–6), and the syllabus design and suggestions for instruction (handling the contents). Examples of the overall objectives of primary science subject together with some of the specific objectives and content in Grade 3 are shown below:

Overall objectives:
To enable pupils to become familiar with nature and to carry out observations and experiments with their own prospectus, as well as to develop their problem-solving abilities and nurture hearts and minds that are filled with an affection for the natural world, and at the same time, to develop a realistic understanding of natural phenomena, and to foster scientific perspectives and ideas

Grade 3

1. Objectives:

(1) To develop perspectives and ideas about the properties and functions of weight, wind, force of rubber, light, and magnets and electricity through investigation comparing phenomena involving these matters, and through probing the identified problem and making learning material with interest.

(2) To foster an attitude of loving and protecting living things and to develop perspectives and ideas about the relationship between living things and the environment, the relationship between the sun and its effects on conditions on earth, through investigation comparing familiar animals and plants, and sunny and shady spots, as well as through probing the identified problems with interest.

2. Content:

A. Matter/Energy

(1) Object and weight

To develop pupils' ideas about properties of objects by examining the weights and volumes, using objects such as clay.

a. The weight of an object remains unchanged even when the shape changes.

b. Objects with the same volume may differ in weight.

B. Life/the Earth

(1) Insects and plants

To develop pupils' ideas about growth patterns and body structures by finding and raising familiar insects and plants, and by exploring the processes of their growth and body structure.

a. Insects grow in accordance with a fixed order of growth, and their body parts consist of the head, thorax, and abdomen.

b. Plants grow in accordance with a fixed order of growth, and their body parts consist of roots, stems, and leaves. (MEXT 2008a)

The learning objectives from every grade belong to either of two parts—Matter/energy and Life/the Earth—taking into account the characteristics of the learning objects, and the perspectives and ideas developed by the pupils. Objectives relating to Matter/energy in each grade include making things/products since the focus here is on *authentic understanding* while those in Life/the Earth include developing an attitude of care and protection of living things, and to respect life since this part focuses on *nurturing hearts and minds that are filled with an affection for the natural world* (MEXT 2008b).

As described in the overall objectives, developing students' problem-solving abilities is strongly emphasized in the primary science curriculum. Problem-solving in the Japanese primary science most often means a series of steps of scientific inquiry. One example of the process of problem-solving consists of "contact with a natural phenomenon," "awareness/questions," "understanding the questions," "assumption/hypothesis," "experiment design," "carrying out the experiment," "obtaining results," "discussion," "conclusion," and "delivery" (MEXT 2006). As abilities or means of problem-solving, the objectives of each grade specify the prioritized problem-solving abilities of the grade, such as comparing phenomena (Grade 3), relating a change with its contributing factors (Grade 4), controlling conditions when observing/experimenting (Grade 5), and reasoning (Grade 6). These abilities are often viewed as *shishitsu noryoku* or competencies in science and some subject matter are intentionally arranged in such a way that the content are well-related to the competencies. This would thus show how the particular competencies in science are useful when solving the given problems (Matsubara 2015).

As seen above, there are only two objectives for Grade 3. Likewise, Grades 4, 5, and 6 each have two objectives, making up only eight objectives for Japan's primary science curriculum. Each objective contains a sizeable amount of information, covering about half of the learning content for the grade. Compared to the learning objectives of five other states, one might say Japan's objectives for each grade tend to be general and broad. Considering such particular circumstances, this study will be dealing with the content of each grade rather than the objectives of each grade in the context of the Japanese primary science curriculum and will be using the descriptions of the content in the analysis of objectives. In this way, we believe we will be able to compare the objectives among the six states with better validity. Two examples of the content of Grade 3 are given above. The descriptions of the content, such as "to develop pupils' ideas about the properties of objects by examining weights and volumes using objects such as clay," are considered to be objectives suitable for analysis.

The content in each grade basically consists of the following points in the following order, to enable pupils to carry out problem-solving activities with learning objects and to achieve the objectives of each grade (MEXT 2008b).

(1) First, the learning object and activities are specified, such as "using objects such as clay" (Another example may be "by finding and raising familiar insects and plants").

(2) Second, the perspective of learning is specified, such as "by examining the weights and volume" (Another example may be "by exploring the processes of their growth and body structure").

(3) Then, the expected ideas to be developed by pupils in the process of their learning or as a result of their learning are specified, such as "develop pupils' ideas about the properties of objects" (Another example may be "develop pupils' ideas about growth patterns and body structures").

(4) Next, descriptions under (a) and (b) show the content of detailed ideas about the learning object, which are expected to be developed by pupils as a result of their learning.

(5) In order to ensure pupils' proactive problem-solving activities, the content is chosen and arranged in a way that pupils can work on natural phenomena and develop scientific perspectives and ideas.

Overall National Profile of Learning Objectives

As we have found from our coding in Table 4.6, there is a distinctive pattern in the learning objectives from Japan that reflects the wording of the official curriculum document to foreground doing activities in order to learn scientific concepts. With such a strong emphasis on the former, it was found that 93.5 % of all items could be categorized in Apply, which implied much hands-on learning of science. This figure as previously mentioned was the highest of any domain (cognitive or knowledge) among all six states that were compared in this book. As well, we found a complete absence of items in Understand or Remember, which normally would have garnered a large percentage of items. None of the typical pairings of Factual-Remember or Conceptual-Understand were present; Procedural-Apply, however, had 25.8 % of all objectives in Japan. Most of the objectives were found in the Conceptual: Apply (67.7 %) pairing followed by Procedural-Apply that was less than half of the former. Compared to other states, there were no items in Evaluate and Create either, but this must be appreciated in the light that the Japanese curriculum was the state here with the fewest number of learning objectives.

Profile of Upper and Lower Primary Learning Objectives

Both Tables 4.7 and 4.8 follow closely the overall national profile, but with relatively few items it is hard to make definitive conclusions about the intellectual demands of primary science as children grow older in Japan.

Table 4.6 Table showing total number of learning objectives ($n = 31$) from Japan classified according to the dimensions of knowledge and cognitive processes in RBT

	Remember	Understand	Apply	Analyze	Evaluate	Create	Number of knowledge items
Factual	0	0	0	0	0	0	0
Conceptual	0	0	21 (91.3) (72.4)	2 (8.7) (100)	0	0	23 (74.2)
Procedural	0	0	8 (100) (27.6)	0	0	0	8 (25.8)
Metacognitive	0	0	0	0	0	0	0
Number of cognitive items	0	0	29 (93.5)	2 (6.5)	0	0	31

Percentages shown in brackets (%)

Table 4.7 Number of learning objectives in the knowledge domain from Japan sorted according to their grade levels

Grade	Factual	Conceptual	Procedural	Metacognitive	Total
3–4	0	12 (80.0)	3 (20.0)	0	15 (48.4)
5–6	0	11 (68.8)	5 (31.2)	0	16 (51.6)

Percentages shown in brackets (%)

Table 4.8 Number of learning objectives in the cognitive domain from Japan sorted according to their grade levels

Grade	Remember	Understand	Apply	Analyze	Evaluate	Create	Total
3–4	0	0	15 (100)	0	0	0	15 (48.4)
5–6	0	0	14 (87.5)	2 (12.5)	0	0	16 (51.6)

Percentages shown in brackets (%)

The People's Republic of China

General Introduction

In the national curriculum for elementary education, there are several subject-based courses such as Chinese literature, mathematics, foreign language (English), ethics and society (ethics and life), science, physical education, music, and arts. Science in primary schools is taught from Grades 3 to 6 and teaching hours vary from place to place, but it is taught at least two hours per week and by teachers who are content generalists. The present primary science curriculum document—*Full-time Compulsory Science Curriculum Standards (Grade 3–6)*—was officially released by the Ministry of Education in mainland China for implementation in 2001. Realizing that science education had formerly paid too much emphasis to the acquisition of basic science knowledge and had largely ignored students' interests or needs (Ding 2001), Chinese science educators and curriculum developers now wanted to cultivate students' scientific literacy through inquiry and scientific processes. The latter, as well as the nurturing of scientific attitudes became key goals of science teaching as described below in the curriculum:

- Students know basic scientific knowledge related to common phenomena in daily life and are able to apply this knowledge in everyday life situations, and gradually develop scientific habits and lifestyles
- Students understand the processes and methods of scientific inquiry, attempt to use them in inquiry activities, and gradually learn to think and solve problems in scientific ways

- Students maintain and develop curiosity about the world and thirst for knowledge, cultivate scientific attitudes (including bold imagination, respect for evidence, and innovation), and emotions (including the love of science and homeland)
- Students are willing to get close to and appreciate nature, cherish life, participate in the protection of natural resources and environment, and be aware of new developments in technology.

Unlike the previous curriculum, knowledge and practical skills no longer formed the main focus; memorization, concept-based tests, and technique-based, hands-on skills of science teaching were to be replaced by evidence-based inquiry as the main teaching approach. The emphasis on scientific inquiry and attitudes was also evident in how the learning standards were worded. For instance, some inquiry skills were distinctly stated as learning objectives such as "students know that in scientific inquiry, asking and answering questions and their outcomes have to be compared with one's own finding as well as conclusions from science and students know that different questions need different methods of inquiry." In all, these were very bold and ambitious goals that subscribed to an underlying constructivist philosophy of learning that was a major and genuine reform as commentators have noted (OECD 2011). However, as seen in other Asian countries (e.g., Kim et al. 2013), inquiry-based instruction in everyday practice in China is challenging because of teachers' emphases on mastering content knowledge, students' heavy workloads, and assessment anxieties in the educational system (Huang and Mao 2013). The aforementioned intentions for holistic student learning that fall under the umbrella of "quality education" (*suzhi jiaoyu*) reforms have been hard to implement evenly or successfully across this vast and populous country (Dello-Iacovo 2009).

Under the overall goals of the curriculum, content standards are listed and grouped into five main categories: (1) scientific inquiry, (2) scientific attitudes and values, (3) life science, (4) physical science, and (5) earth and space science. The content standards in each category are stated in accordance with the level of difficulties, i.e., from easy to hard to achieve. Unlike other countries, the learning standards are not grouped by grade levels: The curriculum developers explained that grouping standards by the level of difficulties rather than grades would provide more space and choices for textbook publishers, and implement the policy of "one standard, multiple versions of textbook" (Zhong 2009). The flexibility and openness of content and learning outcomes for each grade could potentially develop thoughtful and appropriate decision making by teachers in various regions and communities. However, the nature of integration as well as openness of the curriculum structure without specific learning objectives for students' grade levels has been challenging for curriculum developers/implementers to make decisions on cognitive characteristics in the formation and development of scientific literacy (Huang and Mao 2013). Thus, in the draft of a newly recommended science curriculum, the content standards were grouped by grade levels (e.g., Primary Science Curriculum Standard Revision Project Team 2013, 2014).

Science textbooks are the main resource of science teaching and in some rural areas, it may be the only resource available for teachers. Thus, it is critical to know what content, activities, and guidelines for learning are provided in science text-books bearing in mind unique contextual factors. In China, there are three main versions of science textbooks, published by the Education Science Publishing House, People's Education Press and Jiangsu Education Press. In the 2001 curriculum document, there are also suggestions of teaching and learning activities, assessment, development and utilization of curriculum resources, and teachers' professional development to assist teachers and schools in planning and implementing the curriculum. In sum, the science curriculum in China has experienced tremendous changes in terms of the curriculum goals, content matter, and pedagogical approaches over the years (Wei 2010).

Overall National Profile of Learning Objectives

In Table 4.9, the majority of learning objectives from China are clustered around Understand-Conceptual (29.0 %). Like other states, certain knowledge-process combinations (Remember-Factual, Understand-Conceptual, Apply-Procedural) were also observed to be the dominant pattern (those three accounted for 56.2 % of items in the intended curriculum). In the cognitive dimension, Remember (29.6 %) and Understand (38.9 %) accounted for most of the items with relatively fewer in the category of Analyze, Evaluate, and Create. In terms of knowledge, it can be seen that Conceptual was the major category accounting for over half (54.3 %) of the items. Factual and Procedural items held lower percentages, and only a single item is in Metacognitive. Learning standards are not categorized in grade levels, thus, there are no tables of learning objectives sorted according to grade levels for mainland China (see Appendix A).

Table 4.9 Table showing total number of learning objectives ($n = 162$) from mainland China classified according to the dimensions of knowledge and cognitive processes in RBT

	Remember	Understand	Apply	Analyze	Evaluate	Create	Number of knowledge items
Factual	29 (67.4) (60.4)	14 (32.6) (22.2)	0	0	0	0	43 (26.5)
Conceptual	19 (21.6) (39.6)	47 (53.4) (74.6)	17 (19.4) (53.1)	4 (4.5) (66.7)	1 (1.1) (20.0)	0	88 (54.3)
Procedural	0	1 (3.3) (1.6)	15 (50.0) (46.9)	2 (6.7) (33.3)	4 (13.3) (80.0)	8 (26.7) (100)	30 (18.6)
Metacognitive	0	1 (100.0) (1.6)	0	0	0	0	1 (0.6)
Number of cognitive items	48 (29.6)	63 (38.9)	32 (19.8)	6 (3.7)	5 (3.1)	8 (4.9)	162

Percentages shown in brackets (%)

Taiwan

General Introduction

The national primary curriculum in Taiwan consists of eight main learning areas: literature (including Chinese, English, and other local/ethnic language), health and physical education, social studies, arts, math, science, and integrated activities. In parallel with those learning areas, it also includes seven major issues, which are supposed to be embodied in each of those learning areas. The seven major issues are: Gender equality, environmental education, information education, education of household management, human rights education, career development education, and marine education. In other words, by learning the primary science curriculum, students also have the opportunities to explore those major issues.

Even though the latest analysis of the 2008 primary science curriculum in Taiwan was published eight years ago by Chiu (2007), it remains the most comprehensive to date and will be referred to extensively here. Since the mid-1990s, science curriculum development in Taiwan has been very much influenced by policy documents from the United States of America. These influences, nonetheless, are to be subsumed under the two main thrusts of educational reform efforts in Taiwan, which are the improvement of equal and excellent opportunities to education among learners (Peng et al. 2011). Chief among the reform recommendations for improving science teaching then was to emphasize inquiry-based instruction, which struggled (as with so many other states) with superficial implementation of such difficult teaching approaches (Abd-El-Khalick et al. 2004). The curriculum structure consists, for our purposes, of three levels: Grades 1–2 (Level 1, life curriculum), Grades 3–4 (Level 2, integrated science and technology curriculum), and Grades 5–6 (Level 3, integrated science and technology curriculum). The general aims of the science and technology curriculum in Taiwan are as follows:

- Cultivate interest and passion for scientific inquiry and to develop habits of active learning
- Learn the methods of science and technology during inquiry and be able to apply these to present and future life
- Develop care for the environment, to conserve resources, and to respect life
- Be able to communicate, work in teams, and work in harmony
- Develop and stimulate the potential for independent thinking and problem-solving skills
- Be aware of and explore interactions between Man and technology.

There are also eight main topics that comprise scientific literacy in the curriculum and are the following:

1. Process skills
2. Knowledge of science and technology
3. Nature of science and technology
4. Development of science and technology

5. Scientific attitudes
6. Habits of thinking
7. Applications of science
8. Design and production.

Five major topical areas further structure the learning standards or guidelines: Composition of materials and the earth's environment; uses of nature; life and the environment, and sustainability. There are no national exit exams at the end of primary school in Taiwan although it has been reported that adequate coverage of content matter by teachers very anxious of time constraints in a crowded curriculum have been pressing concerns for teachers in the previous decade (Abd-El-Khalick et al. 2004). We note with some regret that while Taiwanese science education researchers are among the most prolific in Asia in terms of their representation in the top international journals, their focus from 1998 to 2007 has predominantly been on empirical teaching-learning research/methods at secondary levels (Tsai and Wu 2010). In-depth studies in English on the curriculum in Taiwan, especially at the elementary grades, have not picked up since that previous review (see also Chen et al. 2010). In Appendix B, we list down the learning objectives (excluding those in the affective domain e.g., [5] Habits of Thinking) from Taiwan.

Overall State Profile of Learning Objectives

In Table 4.10, it is clear that learning objectives from Taiwan in the cognitive domain span the whole range of categories with the majority in Apply (38.6 %) and a very respectable number in Create (10.5 %). Conceptual and Procedural both accounted for 82.5 % of items in the knowledge domain. And if the first three knowledge-process combinations are all added together, they account for 71.1 % of the learning objectives from Taiwan.

The pairing of Procedural-Apply moreover garnered nearly a third of all Taiwanese learning objectives, which points to the extremely strong emphasis on achieving understanding of scientific concepts and skills through practical activities. Lu and Lien (2016) have confirmed this inherent bias towards laboratory work when they reported that teachers' responses on the following questions (from TIMSS 2011) concerning the frequency of activities in local classrooms were above international averages (in the USA, the Russian Federation, Korea, Singapore, and Japan):

- Observe natural phenomena such as the weather or a plant growing and describe what they see
- Watch me[the teacher] demonstrate an experiment or investigation
- Design or plan experiments or investigations
- Conduct experiments or investigations.

Table 4.10 Table showing total number of learning objectives ($n = 114$) from Taiwan classified according to the dimensions of knowledge and cognitive processes in RBT

	Remember	Understand	Apply	Analyze	Evaluate	Create	Number of knowledge items
Factual	19 (100) (100)	0	0	0	0	0	19 (16.7)
Conceptual	0	27 (67.5) (96.4)	9 (22.5) (20.5)	2 (5.0) (33.3)	0	2 (5.0) (16.7)	38 (35.1)
Procedural	0	1 (1.8) (3.6)	35 (64.8) (79.5)	4 (7.4) (66.7)	5 (9.3) (100)	9 (16.7) (75.0)	54 (47.4)
Metacognitive	0	0	0	0	0	1 (100) (8.3)	1 (0.9)
Number of cognitive items	19 (16.7)	28 (24.6)	44 (38.6)	6 (5.3)	5 (4.3)	12 (10.5)	114

Percentages shown in brackets (%)

On the other hand, the frequency of responses from Taiwanese teachers to the following questions were the lowest in their international comparative study:

- Read their textbooks or other resource materials
- Have students memorize facts and principles
- Give explanations about something they are studying
- Relate what they are learning in science to their daily lives
- Take a written test or quiz.

Profile of Upper and Lower Primary Learning Objectives

As mentioned, there was a clear emphasis on Procedural (and its concrete manifestation through practical work) in the Taiwanese curriculum. Hence, it comes as no surprise that in Table 4.11 we observe that across all grade divisions this category predominated, with Conceptual equal or a close second in terms of representation (other than Grades 3–4). Conversely, items in Factual were not as

Table 4.11 Number of learning objectives in the knowledge domain from Taiwan sorted according to their grade levels

Grade	Factual	Conceptual	Procedural	Metacognitive	Total ($n = 114$)
1–2	1 (4.8)	10 (47.6)	10 (47.6)	0	21 (18.4)
3–4	8 (25.0)	7 (21.9)	17 (53.1)	0	32 (28.1)
5–6	10 (16.4)	23 (37.7)	27 (44.3)	1 (1.6)	61 (53.5)

Percentages shown in brackets (%)

Table 4.12 Number of learning objectives in the cognitive domain from Taiwan sorted according to their grade levels

Grade	Remember	Understand	Apply	Analyze	Evaluate	Create	Total (n = 114)
1–2	1 (4.8)	5 (23.8)	13 (61.8)	1 (4.8)	0	1 (4.8)	21 (18.4)
3–4	8 (25.0)	4 (12.5)	12 (37.4)	2 (6.3)	2 (6.3)	4 (12.5)	32 (28.1)
5–6	10 (16.5)	19 (31.1)	19 (31.1)	3 (4.9)	3 (4.9)	7 (11.5)	61 (53.5)

Percentages shown in brackets (%)

numerous other than when in Grades 3–4. As children progress in primary school, it is to be noted that the number of science learning objectives increases from 21 to 32 to 61, a threefold increase from the earlier years.

The trend toward practical knowing in science was likewise reflected in Table 4.12 for the number of items in Apply accounted for the vast majority of objectives across two of the three divisions while they were tied with items in Understand in Grades 5–6. We observe too that items in Create were represented at all divisions though more so in the latter two. Finally, it was reported by Lu and Lien (2016) that the content distribution of Grade 4 textbooks in Taiwan were biased toward physical science topics; percentages of topics on biology, physical science, and life sciences were 29.8, 39.1, and 18.5 % compared to international averages of 38.1, 25.3, and 23.1 % respectively.

The Republic of Korea

General Introduction

The 2015 science curriculum aims to develop students' scientific literacy to solve individual and social problems scientifically and creatively. This was to be accomplished through understanding essential scientific concepts and developing inquiry skills (MOE 2015). The specific goals stated in this new curriculum are as follows:

1. Students cultivate attitudes of curiosity toward scientific problem-solving and have interest in natural phenomena
2. Students develop abilities to scientifically inquire about natural phenomena and everyday problems
3. Students understand essential scientific concepts through inquiry about natural phenomena

4. Students understand the interrelationships among science, technology, and society and develop citizenry literacy in a democratic society based on this understanding
5. Students cultivate lifelong learning abilities through recognizing the joy of science learning and application of science.

Science is taught from K-12 in Korean school systems but science as a separate subject is taught only from Grades 3 to 12. This means that there are no specific science units or learning objectives stated for K-2 levels in the national science curriculum. In Grades 1–2, subjects include the Korean language, Mathematics, and three integrated subjects: Intelligent Life, Disciplined Life, and Pleasant Life. Science is included under the subject of Intelligent Life. From Grades 3–9, science is taught as an independent subject and later in Grade 10, Integrated Science and Science Inquiry Experiments are also included. In Grades 11–12, science is divided into more specific disciplines such as physics, chemistry, life science, and earth and space science, history of science, science and everyday life, and interdisciplinary science. Thus, in primary schools, children in Grades 1–2 learn science in the Wise life subject and only in Grades 3–6 then they learn science as an independent school subject.

The primary science curriculum has four overall dimensions; Movement and energy, Matter, Life, and Earth and space. Each dimension includes several topics. The science topics in Grades 3–4 include the characteristics of materials and matters, usage of magnets, lives of animals, changes of the earth surface, lives of plants, strata and fossils, states of matters, characteristics of sound, weight, life-cycles of animals, volcanoes and earthquakes, separation of mixtures, lifecycles of plants, changes of states of water, shadows and mirrors, shape of the earth, and the movement of water (integrated topic). Science in Grades 5–6 includes temperature and heat, the solar system and stars, dissolution and solution, various living beings and our lives, life and the environment, weather and our lives, movements of objects, acids and bases, movements of the earth and moon, various gases, light and lenses, structures and functions of plants, usage of electricity, seasonal changes, combustion and fire extinction, structures and functions of our bodies, and energy and everyday lives (integrated topic). The two integrated topics are newly included in 2015 science curriculum to develop integrated science learning and knowledge application.

The topics and units in the primary science curriculum are developed in consideration of the level of children's cognitive development and interconnections throughout the grade levels and it is explained in the science curriculum. Here is one example (Table 4.13) of knowledge connection on electricity and magnetism from primary Grades 3–6 to middle year Grades 1–3.

In the 2015 science curriculum (see Appendix C), each unit introduces specific learning objectives and inquiry activities, which are part of mandatory elements for teaching. It is a unique feature of Korean curriculum that these inquiry activities are stated in connection with learning objectives to emphasize the importance of inquiry teaching. For instance, the first unit in Grade 3 science is Characteristics of

Table 4.13 An example of the connection of primary science concepts across the grade levels in Korea (MOE 2015, p. 5)

Dimension	Core concept	General knowledge	Content elements		
			Primary		Middle
			G 3–4	G 5–6	G 1–3
Electricity and Magnetism	Electricity	There is electric force between two electric charges			– Electric force – Atomic model – Electrification – Electrostatic induction
		Current is formed by electromotive force in electric circuits		– Electric circuit – Electricity conservation – Electricity safety	– Electric circuit – Voltage – Current – Resistance
	Magnetism	Current forms magnetic field		– Electromagnet	– Magnetic field – Electromotor – Electricity generation
		Matters can be classified as magnetic and nonmagnetic	– Magnetic force – Characteristics of magnets		

materials and matters. In this unit, four learning objectives and three inquiry activities are introduced as shown below.

Learning objectives:

1. By comparing objects made of different materials and matters, students can make connections between functions and characteristics of objects.
2. Compare various characteristics of materials and matters by observing objects with the same shapes and sizes but different materials and matters.
3. Explain the changes of characteristics of matters by observing the differences made between before and after different matters are mixed.
4. Design various objects with various materials and matters and discuss their strengths and weakness.

Inquiry activities:

- Investigate what matters and materials objects are made of
- Relate the functions and characteristics of objects
- Observe changes in characteristics of materials and matters.

There is thus a brief instruction on the interpretation of learning objectives, instructional guide, and assessment guide in the science curriculum but the details of those ideas are explained in teachers' guidebooks which are to be developed much later after the curriculum is released. During this period of curriculum reform, a team of science teacher educators, primary science teachers, and scientists develop the science textbooks and teacher guidebooks. This team is assembled by the government and the science textbooks and teacher guidebooks developed by this team become the official resources that all primary school teachers use as a working version of science curriculum across the nation. The mandatory teaching hours of science is about 102 h per year. In most of primary schools, science is taught by homeroom teachers in Korea as normally seen in other countries.

Overall National Profile of Learning Objectives

Table 4.14 shows that 79.6 % of learning objectives belong to Conceptual and 19.5 % are in Procedural. In terms of cognitive processes, 62 % are located in Apply, 34.5 % in Understand, and 3.5 % in Create. In the Korean curriculum, there are more items in the Conceptual:Apply pair ($n = 51$) than in Conceptual: Understand ($n = 39$) and Procedural:Apply ($n = 18$). This can be explained by how the learning objectives in the new 2015 science curriculum were structured; many learning objectives require students to conduct activities to achieve the understanding of scientific concepts. For example, one of the learning objectives in Grade 3–4, "By comparing objects made of different materials and matters, students can make connections between functions and characteristics of objects," students have to compare objects, substances, and materials (Apply) in order to understand the relationships between functions and characteristics of the objects (conceptual knowledge). In the latest science curriculum reform "by doing," i.e., students' actions are emphasized more than in the previous version of science curriculum in 2009, thus, the proportion of Apply-Conceptual pairing are more frequent than the Understand-Conceptual one (Lee et al. 2015; Na et al. 2015). Compared to the

Table 4.14 Table showing total number of learning objectives ($n = 113$) from Korea classified according to the dimensions of knowledge and cognitive processes in RBT

	Remember	Understand	Apply	Analyze	Evaluate	Create	Number of knowledge items
Factual	0	0	1 (100.0) (1.4)	0	0	0	1 (0.9)
Conceptual	0	39 (43.3) (100.0)	51 (56.7) (72.9)	0	0	0	90 (79.6)
Procedural	0	0	18 (81.8) (25.7)	0	0	4 (18.2) (100.0)	22 (19.5)
Metacognitive	0	0	0	0	0	0	0
Number of cognitive items	0	39 (34.5)	70 (62.0)	0	0	4 (3.5)	113

Percentages shown in brackets (%)

Table 4.15 Number of learning objectives in the knowledge domain from Korea sorted according to their grade levels

Grade	Factual	Conceptual	Procedural	Metacognitive	Total (n = 113)
3–4	0	45 (78.9)	12 (21.1)	0	57 (50.4)
5–6	1 (1.8)	45 (80.3)	10 (17.9)	0	56 (49.6)

Percentages shown in brackets (%)

Table 4.16 Number of learning objectives in the cognitive domain from Korea sorted according to their grade levels

Grade	Remember	Understand	Apply	Analyze	Evaluate	Create	Total (n = 113)
3–4	0	19 (33.3)	35 (61.4)	0	0	3 (5.3)	57 (50.4)
5–6	0	20 (35.7)	35 (62.5)	0	0	1 (1.8)	56 (49.6)

Percentages shown in brackets (%)

previous curriculum, factual knowledge was significantly reduced. Overall, it is clear that Conceptual and Apply are emphasized in the recent science curriculum 2015.

Profile of Upper and Lower Primary Learning Objectives

The number of learning objectives for Grades 3–4 and Grades 5–6 are almost equal (n = 57 in Grades 3–4, n = 56 in Grades 5–6). In the span of the knowledge dimension and cognitive processes in lower and upper grade levels, there are few differences in the Korean curriculum (see Tables 4.15 and 4.16). In terms of knowledge, 78.9 % of items belong to Conceptual in Grades 3–4 and 80.3 % in Grades 5–6. There is only one learning objective in Factual in Grades 5–6 with none in Metacognitive. For cognitive processes, Understand takes up about 20 % and Apply about 62 % of all the learning objectives. There are three learning objectives in Create in Grades 3–4 and one in Grades 5–6.

Singapore

General Introduction

The latest primary science syllabus in Singapore took effect in 2014 (see CPDD 2013) with a renewed emphasis on teaching and learning science via inquiry. Inquiry was the unifying idea in a broad framework that included knowing science in daily life, in society, and the environment as shown in Fig. 4.1.

Fig. 4.1 The science
curriculum framework for
primary schools in Singapore.
(The Government of
Singapore (c/o Ministry of
Education) owns the
copyright to the figure and the
figure is reproduced with their
permission.)

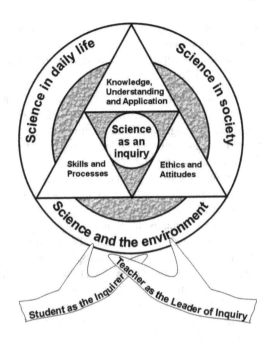

Knowing subject matter here is understood in terms of two aspects (i) knowledge, and (ii) understanding with application: the former is the more fundamental
and can be considered analogous to the first level of the knowledge domain in RBT
while the latter finds affinities with the second and third levels in RBT. In common
with many other countries, the learning of skills and process in science is also felt to
be fundamental together with cultivating wholesome ethics and attitudes in science.
This curriculum document further associate the learning of science with a number
of so-called twenty-first century competencies: A flexible set of skills and values as
the "outcomes of individuals able to thrive in and contribute to a world where
change is the only constant" (CPDD 2013, p. 3). Another important goal of learning
science in Singapore is the achievement of scientific literacy, which was described
here using words adapted from the PISA 2006 science framework. No longer was
there a fixation on merely memorization of more and more facts, but instead
evidence-based reasoning, and application of knowledge to make decisions in the
real world are seen as highly desirable.

The syllabus is very specific in spelling out the overall aims of learning primary
science that mirror the science curriculum framework:

- provide students with experiences which build on their interest in and stimulate
  their curiosity about their environment
- provide students with basic scientific terms and concepts to help them understand themselves and the world around them
- provide students with opportunities to develop skills, habits of mind, and attitudes necessary for scientific inquiry

- prepare students toward using scientific knowledge and methods in making personal decisions
- help students appreciate how science influences people and the environment (see CPDD 2013, p. 5).

In terms of the organization of science content, the learning outcomes are grouped into five integrated themes (Diversity, Interactions, Systems, Cycles, Energy). Although somewhat arbitrary, these groupings were thought to be able to "communicate a more coherent and integrated understanding of science that bridged the life science-physical science divide" (Chin and Poon 2014, p. 33). Again, the agency of students is underscored as they are supposed to be critical inquirers in science under the guidance of the teacher who plays the role of the leader or facilitator of inquiry (Fig. 4.1).

To assist teachers in planning and implementing inquiry-based lessons, the syllabus suggests a useful scaffold for inquiry—the five essential features that originated first in the Biological Sciences Curriculum Study (BSCS) (Loucks-Horsley and Olson 2000). Within this framework, it allows educators to plan a spectrum of teacher- and student-centered modes of teaching. In this state, the syllabus is divided into two blocks: lower- (Grades 3–4) and upper-primary (Grades 5–6). Due of the presence of exit examinations (in four subjects including science) at the end of Grade 6 in Singapore, many researchers have reported that the intended curriculum was largely isomorphic with the taught and tested curriculum (Hogan et al. 2013). The grammar of schooling, not just in science, means that local teachers are very sensitive to assessment concerns despite their familiarity and personal beliefs about teaching science via inquiry.

Overall National Profile of Learning Objectives

According to Table 4.17, the majority of learning objectives from Singapore clustered around Understand-Conceptual (48.3 %) in alignment with most other school curricula (see Appendix D). Certain knowledge-process combinations (Remember-Factual, Understand-Conceptual, Apply-Procedural) were observed to be the dominant pattern here as well; they accounted for 65.1 % of items in the intended curriculum. It is interesting to note that no item appeared beyond Analyze nor were there any representations in Metacognitive. An earlier coding of items in Lee et al. (2015) showed that Understand garnered 60.2 % of all learning objectives while those in Remember were just 13.3 %. Moreover, these authors reported that items in Procedural were about 10 % higher and those that fell in Conceptual were lower by the same amount. We believe that the current Table 4.17 is a more accurate portrayal of the spread of objectives from Singapore; we have improved our coding abilities using RBT that is consistent across all six states (not just two) surveyed in this book.

Table 4.17 Table showing total number of learning objectives ($n = 83$) from Singapore classified according to the dimensions of knowledge and cognitive processes in RBT

	Remember	Understand	Apply	Analyze	Evaluate	Create	Number of knowledge items
Factual	14 (100) (53.8)	0	0	0	0	0	14 (16.9)
Conceptual	12 (21.4) (46.2)	27 (48.2) (100)	13 (23.2) (50.0)	4 (7.2) (100)	0	0	56 (67.5)
Procedural	0	0	13 (100.0) (50.0)	0	0	0	13 (15.6)
Metacognitive	0	0	0	0	0	0	0
Number of cognitive items	26 (31.4)	27 (32.5)	26 (31.4)	4 (4.7)	0	0	83

Percentages shown in brackets (%)

Profile of Upper and Lower Primary Learning Objectives

From Tables 4.18 and 4.19, we see that as children move from lower to upper-primary levels, the number of objectives increased by 13.1 % in Conceptual mainly at the expense of Factual items. With respect to cognitive processes, however, there was little change for most processes over this same period. Science teaching in lower primary occupies about 90 min per week that increases to about 150 min in upper-primary, which can therefore account for the 59 % increase (from

Table 4.18 Number of learning objectives in the knowledge domain from Singapore sorted according to their grade levels

Grade	Factual	Conceptual	Procedural	Metacognitive	Total
3–4	8 (25.0)	19 (59.4)	5 (15.6)	0	32 (38.6)
5–6	6 (11.8)	37 (72.5)	8 (15.7)	0	51 (61.4)

Percentages shown in brackets (%)

Table 4.19 Number of learning objectives in the cognitive domain from Singapore sorted according to their grade levels

Grade	Remember	Understand	Apply	Analyze	Evaluate	Create	Total
3–4	10 (31.3)	9 (28.1)	11 (34.4)	2 (6.2)	0	0	32 (38.6)
5–6	16 (31.3)	18 (35.3)	15 (29.4)	2 (4.0)	0	0	51 (61.4)

Percentages shown in brackets (%)

32 to 51) in learning objectives during this period. Note again our warnings against simple counting of the changes in the numbers of objectives earlier; there are good rationales for these frequencies as children move up grade levels.

References

Abd-El-Khalick, F., BouJaoude, S., Duschl, R., Lederman, N., Momlok-Naaman, R., Hofstein, A., et al. (2004). Inquiry in science education: International perspectives. *Science Education, 88*, 397–419.

Alexander, R. J. (2008). *Essays on pedagogy*. London: Routledge.

CDC [Curriculum Development Council]. (2002). *Science education: Key learning area curriculum guide (Primary 1-Secondary 3)*. Hong Kong SAR: The Education Bureau.

CDC. (2011). *General studies for primary schools curriculum guide (Primary 1-Primary 6)*. Hong Kong SAR: The Education Bureau.

Chen, Y.-T., Liang, J.-C., Lai, M.-L., & Chang, F.-Y. (2010). Science education for young children in Taiwan. In Y.-J. Lee (Ed.), *The world of science education: Handbook of research in Asia* (pp. 121–134). Rotterdam: Sense Publishers.

Chin, T.-Y., & Poon, C. L. (2014). Design and implementation of the national primary science curriculum: A partnership approach in Singapore. In A.-L. Tan, C. L. Poon, & S. S. L. Lim (Eds.), *Inquiry into the Singapore science classroom: Research and practices* (pp. 27–46). Dordrecht: Springer.

CPDD [Curriculum Planning & Development Division]. (2013). *Science syllabus primary 2014*. CPDD: Ministry of Education, Singapore. Retrieved from https://www.moe.gov.sg/docs/default-source/document/education/syllabuses/sciences/files/science-primary-2014.pdf

Chiu, M.-H. (2007). Standards for science education in Taiwan. In D. Waddington, P. Nentwig, & S. Schanze (Eds.), *Making it comparable: Standards in science education* (pp. 303–346). Munster, Germany: Waxmann.

Dello-Iacovo, B. (2009). Curriculum reform and 'quality education' in China: An overview. *International Journal of Educational Development, 29*, 241–249.

Ding, B. (2001). Reflection on science education. *Journal of the Chinese Society of Education, 02*, 21–25. [in Chinese].

Hogan, D., Chan, M., Rahim, R., Kwek, D., Khin, M. A., … Luo, A. (2013). Assessment and the logic of instructional practice in Secondary 3 English and mathematics classrooms in Singapore. *Review of Education, 1*, 57–106. doi:10.1002/rev3.3002

Huang, X., & Mao, C. (2013). The integrated science curriculum in mainland China. In E. H-F. Law & C. Li (Eds.). *Curriculum innovations in changing Societies: Chinese perspectives from Hong Kong, Taiwan and mainland China* (pp. 189–215). Rotterdam, The Netherlands: Sense Publishers.

Kennedy, K. J. (2005). *Changing schools for changing times: New directions for the school curriculum in Hong Kong*. Hong Kong: Chinese University Press.

Kim, M., Tan, A.-L., & Talaue, F. (2013). New vision and challenges in inquiry-based curriculum change in Singapore. *International Journal of Science Education., 35*, 289–311.

Lee, Y.-J., Kim, M., & Yoon, H.-G. (2015). The intellectual demands of the intended primary science curriculum in Korea and Singapore: An analysis based on revised Bloom's taxonomy. *International Journal of Science Education, 37*, 2193–2213.

Loucks-Horsley, S., & Olson, S. (Eds.). (2000). *Inquiry and the national science education standards: A guide for teaching and learning*. Washington, DC: National Academies Press.

Lu, Y.-L., & Lien, C.-J. (2016). Elementary science education in Taiwan—From the perspective of international comparison. In M.-H. Chiu (Ed.), *Science education research and practices in Taiwan: Challenges and opportunities* (pp. 163–180). Singapore: Springer.

Marsh, C., & Lee, C.-K. J. (Eds.). (2014). *Asia's high performing education systems: The case of Hong Kong*. New York: Routledge.

Matsubara, K. (2015). Relationship between competencies and learning content in science: creating scenes in lessons to enable pupils to realize the necessity of competencies. *Elementary School Science Education, Nihon Shotorika Kyoiku Kenkyukai (Japan Elementary Science Education Association), 45*(3), 7–10 (in Japanese).

MEXT [Ministry of Education, Culture, Sports, Science & Technology]. (2006). *Images of Revising the Learning Content of Science (Draft)*, Central Council for Education, Primary and Secondary Education Sectional Committee, Curriculum Subcommittee (fourth stage, 11th) conference minutes/documents distributed 4–2.

MEXT (2008a). *Shogakko gakushu shido yoryo [Course of Study for Elementary Schools]*, Tokyo Shoseki [In Japanese]. An English translation (tentative version) is available on the ministry's website (http://www.mext.go.jp/component/a_menu/education/micro_detail/__icsFiles/afieldfile/2009/04/21/1261037_5.pdf).

MEXT (2008b). *Shogakou gakushu shido yoryo kaisetsu rika-hen [Explanations for the Course of Study for Elementary Schools: Science]*, Dainippon Tosho [In Japanese].

Ministry of Education, Korea. (2015). *Science curriculum*. Seoul, Korea: Ministry of Education. (in Korean).

Na, J., Yoon, H.-G., & Kim, M. (2015). Analysis of the alignment between elementary science curriculum and teacher guidebook: Examining learning objectives in 2009 Grade 3–4 science curriculum. *Journal of Korean Elementary Science Education, 34*(2), 183–193.

OECD [Organisation for Economic Co-operation and Development]. (2011). *Strong performers and successful reformers in education: Lessons from PISA for the United States*. Paris: OECD Publishing.

Peng, S. S., Yeh, C.-L., & Lee, C.-K. J. (2011). Educational reforms and school improvement in Taiwan. In J. C.-K. Lee & B. J. Caldwell (Eds.), *Changing schools in an era of globalization* (pp. 40–66). New York: Routledge.

Primary Science Curriculum Standard Revision Project Team. (2013). *Full-time compulsory science curriculum standards (Grade 1–6)* (Unpublished Revised Version). Retrieved from http://www.doczj.com/doc/65485098ba0d4a7303763a13-4.html

Primary Science Curriculum Standard Revision Project Team. (2014). *Full-time compulsory science curriculum standards (Grade 1–6)* (Unpublished Revised Version). Retrieved from http://www.doczj.com/doc/32a61b1e2e3f5727a5e962e7-3.html

Tsai, C.-C., & Wu, Y.-T. (2010). Science education research in Taiwan. In Y.-J. Lee (Ed.), *The world of science education: Handbook of research in Asia* (pp. 35–50). Rotterdam: Sense Publishers.

Wei, B. (2010). The changes in science curricula in China after 1976: A reflective review. In Y.-J. Lee (Ed.), *The world of science education: Handbook of research in Asia* (pp. 89–102). Rotterdam: Sense Publishers.

Zhong, Q. (2009). "One Standard, multiple versions of textbook" pursuit of education democracy: A review of textbook policy in China. *Exploring Education Development, 04*, 01–04. (in Chinese).

Chapter 5
The Shape of Intellectual Demands in East-Asian Primary Science Curricula

What can we learn from this study that explored the intellectual demands of the intended primary science curriculum among six East-Asian states? First of all, we do not detect patterns in any state that showed that their learning objectives were skewed towards the higher-order categories in RBT. For all states with the exception of Japan, we find a general distribution of items that parallel those from many other curricula in the world (e.g., DeMers 2009; Fitzpatrick and Schulz 2015). That is, there was a clustering of items within the first three categories within the knowledge as well as the cognitive dimensions of RBT. Conversely, there was a scarcity of items beyond Analyze (ranging from 0 to 10 %) (Tables 4.1 and 4.2 in Chap. 4) just as there were very few objectives located in the metacognitive category (from 0 to 2 %). There seems to be some cause for further research into whether East-Asian students are exposed to sufficient challenge as their governments are eager to devise curricula that raise both the quality and quantity of scientific literacy among school-going populations. Recall, however, that these results might possibly be an effect of adopting RBT that was originally designed by educational psychologists rather than a true reflection of the intellectual work needed to do, understand, and apply science in school. Nor do these findings derived from our coding practices have any close relationship with what goes on in classrooms in these states. That is, excellent science instruction can occur despite supposedly low figures in RBT in the intended curriculum.

Similarly, we have found a high frequency of the typical pairings that other researchers using RBT to analyze school curricula have reported: Factual-Remember, Conceptual-Understand, and Procedural-Apply. In this sense, all of their curricula profiles in RBT have not deviated much from each other or from other jurisdictions around the world and is perhaps to be expected for guiding age-appropriate primary science instruction. Although it might seem that the Japanese curriculum has a peculiar configuration, we claim that it mirrors the intentions and more so the phrasing of its curriculum developers that have led to this profile according to RBT.

© The Author(s) 2017
Y.-J. Lee et al., *East-Asian Primary Science Curricula*,
SpringerBriefs in Education, DOI 10.1007/978-981-10-2690-4_5

As we have explained at the beginning of this book, it is certainly impossible to make any assertions from the data that one educational system has a "better" curriculum simply by examining the intended primary science curriculum in this manner. Apart from reasons concerning the generic nature of coding for cognition in RBT or unintentional classification errors, school curricula are what they are because of unique sociopolitical factors and layers of educational philosophies that have been laminated on them over many years. If anything, we have discovered a surprising diversity in how national curricula in primary science have been organized in East-Asia: States such as Hong Kong have embedded science learning objectives within a general studies/integrated curriculum whereas in other states science is a stand-alone subject. Yet, science begins from Grade 1 in Taiwan but in the rest of the four states it starts two years later. Among all the states examined here, only students from Singapore will undertake an exit exam for science at the end of their primary education. All these structural factors will differently influence how science is taught, assessed and valued by young children in each state.

The number of primary science learning objectives in the cognitive domain that varies from 31 in Japan to 162 in the case of China strikingly bears no correlation with the quality of student achievement in science as reported in the international standardized tests. It is thus most gratifying to learn that more is not necessarily better with regard to how much a child has to master before leaving primary school. Observant readers will also notice the prevalence of what we call "nested" learning objectives in some states (e.g., China, Taiwan, Japan, Singapore) whereby a number of ideas or concepts are described within a single objective. For consistency in our coding, these sub-points were not separated and indeed, they are regarded as a singularity by local teachers which again highlights the danger of making judgments about merit or difficulty based on mere counting of curriculum objectives. Some states such as Japan and Korea have also placed a heavy emphasis on learning through doing science while others express learning in more traditional terms regarding the nature and learning of scientific concepts. Such differences in conceptions of how good science teaching ought to proceed thus reflect cultural-historical nuances and priorities about instruction. Despite these and many other differences, their curricula profiles do appear quite similar (and mundane) when coded by RBT, which we have earlier suggested might be due to policy borrowing and globalization.

What our research has not delved into any extent is a comparison of the coverage (or exclusion) and emphasis of various science topics among the six states (see TIMSS, 2007). For example, some states such as China and Taiwan have detailed and numerous standards related to learning about the Nature of Science and doing experiments (see Appendices A to D). Such meta-knowledge about performing investigations is, however, largely absent from other states such as Singapore or Japan. At least from our observations, Singapore seems to devote the largest number of learning objectives towards learning about the topic of reproduction compared to the other states. Again, this fact must be considered in the light that aspects of health and personal safety might be harboring under the ambit of other school subjects. Whether earth science learning objectives are considered part of science or a separate subject as in geography or even to be reserved for secondary schooling is likewise moot; different states have arrived at different (legitimate) conclusions (Lee et al. 2015). All these specifics about what and how many scientific ideas elementary school children will have opportunities to acquire are arguably some of the most significant questions to inquire if policymakers wish to understand and improve scientific literacy. It is clear that this will be a long-term project that has been made more achievable now given our new English translations of their learning objectives that have previously been published only in the vernacular. We are optimistic in anticipating many more comparative studies in primary science in East Asia and the rest of the world.

References

DeMers, M. N. (2009). Using intended learning objectives to assess curriculum materials: The UGIS body of knowledge. *Journal of Geography in Higher Education, 33*(Supplement 1), S70–S77.

Fitzpatrick, B., & Schulz, H. (2015). Do curriculum outcomes and assessment activities in science encourage higher order thinking? *Canadian Journal of Science, Mathematics and Technology Education, 15,* 136–154.

Lee, Y.-J., Kim, M., & Yoon, H.-G. (2015). The intellectual demands of the intended primary science curriculum in Korea and Singapore: An analysis based on revised Bloom's taxonomy. *International Journal of Science Education, 37,* 2193–2213.

TIMSS (2007). Appendix C. The test-curriculum matching analysis: Science. Available http://timss.bc.edu/timss2007/PDF/T07_S_IR_AppendixC.pdf

Appendix A
Primary Science Learning Standards (Cognitive Domain) from China Based on Our English Translation

Learning objectives (our translation)
1. Know that in scientific inquiry, asking and answering questions, these have to be compared with one's own findings as well as conclusions from science
2. Know that different questions need different methods of inquiry
3. Know why using instruments is more effective than relying on the senses
4. Understand that the results of scientific inquiry should be open to re-verification
5. Know that questioning the results of other investigations is part of scientific inquiry, understand that having reasonable doubt motivates the progress of science
6. Know that exchanging (ideas) and having discussions can lead to new ideas
7. Know that scientific inquiry can provide new experiences, new phenomena, new methods, and new technologies to advance research
8. Be able to raise questions about local events from different perspectives such as "what is this?" or "why is it so/why does this happen?"
9. Be able to select appropriate questions for one's own inquiry
10. Be able to compare and evaluate questions that have been raised
11. Be able to hypothesize explanations for observed phenomenon based on existing knowledge and experiences
12. Be able to distinguish between hypothesis and fact
13. Be able to propose a general outline for conducting inquiry
14. Be able to form a written plan for inquiry raised by oneself or others
15. Be able to perceive natural objects using a variety of senses and be able to describe their external characteristics using words or pictures
16. Be able to perform careful observations using simple tools (magnifier, microscopes, etc.) and express (the results) with diagrams and words
17. Be able to make quantitative measurements of objects and collect data using simple tools (ruler, dynamometer, scale, graduated measuring cylinder, thermometer, stopwatch, etc.), and make simple records
18. Be able to perform simple observational experiments with simple equipment, and make records of experiments
19. Be able to perform simple inquiry experiments with control of variables, design simple reports of experiments, and draw simple charts

(continued)

© The Author(s) 2017
Y.-J. Lee et al., *East-Asian Primary Science Curricula*,
SpringerBriefs in Education, DOI 10.1007/978-981-10-2690-4

(continued)

Learning objectives (our translation)
20. Be able to make simple scientific models (rock samples, insect models, volcanos, stratigraphic models etc.)
21. Be able to integrate knowledge and make objects of technology (crane model, terrarium, energy conversion device etc.)
22. Be able to consult books and other information sources
23. Be able to organize information using tables, graphs, statistics, and other simple methods
24. Be able to analyze, interpret data, and reasonably explain a phenomenon in different ways
25. Be able to consider making different interpretations of the same phenomenon
26. Be able to reflect on the process of one's inquiry and compare the results of inquiry with the hypothesis
27. Be able to choose the method(s) (language, text, graphics, models etc.) which one is good at to express the research process and results
28. Be able to evaluate(or make comments to) the research process and results, and exchange perspectives with others
29. Be able to name common plants in the vicinity and classify these common plants simply
30. Understand local plants as resources, and be able to realize the close relationship between plants and human life
31. Understand more types of plants and experience the diversity of plants
32. Know the names of common animals found in daily life, and be able to classify these animals according to different criteria
33. Make generalizations about the common features of a certain type of animal
34. Recognize several types of common animals, such as insects, fish , amphibians, reptiles, birds, and mammals
35. Understand more types of animals and experience the diversity of the animal world
36. Recognize the diversity of animal movement
37. Understand the importance of protecting animals, especially endangered animals
38. Understand the main characteristics of bacteria and their positive and negative effects on Man
39. Know that fungi belong to a class of organism which is neither plant nor animal
40. Understand that viruses are also a type of organism
41. Understand the process of plant growth through the process of growing plants
42. Describe the general process of animal development through the process of raising small animals
43. Understand that different organisms have different life processes, and experience the complexity and diversity of different life processes
44. Know that reproduction is a common characteristic of life
45. List the different modes of reproduction among commonly seen animals
46. Be able to identify the six major organs of plants, and know the functions of these various organs
47. Investigate the functions of roots and stems
48. Through observations of mammals, specimens or viewing multimedia software, identify some of their major organs
49. Know that cells are the basic units in the living body

(continued)

(continued)

Learning objectives (our translation)
50. Recognize that organisms need to absorb water and nutrients from the outside world to survive
51. Design experiments to study the effects of water, sunlight, air, temperature, and fertilizer etc. on plant growth
52. Know that different animals eat different foods, and animals need the energy from food to survive
53. Understand that green plants can produce starch and oxygen in sunlight, and at the same time absorb carbon dioxide
54. Know that many biological characteristics are hereditary
55. Understand that heredity and variation are one of the major characteristics of organisms
56. Observe the appearance of plants, and make links between the results of these observations and their living environment
57. Be able to carefully observe the external features of animals, and link the results of these observation with their characteristics
58. List some specific examples of adaptations to the environment among similar types of organisms
59. Understand some properties of plants that respond to the environment, such as phototropism, hydrotropism, and geotropism
60. List some examples of how animals adapt to the environment, such as hibernation, camouflage, mimicry, etc.
61. Know that the environment has influences on biological growth and many other aspects of life
62. Know the meaning of a food chain
63. Recognize that human beings are part of nature, they not only depend on the environment, but also affect the environment, and can influence the survival of other creatures
64. Be able to explain the meaning of survival of the fittest and natural selection
65. Be able to explain the process of biological evolution, with specific organisms as examples
66. Understand what are the needs and nutritional sources for human beings, and understand the importance of complete and proper nutrition
67. Understand the process of human digestion and develop good habits of eating, drinking, and health
68. Understand the process of human respiration, and know the origin and prevention of common respiratory diseases
69. Understand the functions of heart and blood vessels and how to take care of them
70. Investigate what factors affect the rate of heartbeat
71. Understand the function of sensory organs and know that the body's various senses are responses to the outside environment
72. Know the role of the brain in language, thinking, emotions, and that it is the "command center" of human life processes
73. Understand the general process of growth across a person's lifespan
74. Understand the special characteristics of physical development among youth
75. Understand the special characteristics of physical and mental development in adolescents
76. Understand the various factors that affect health

(continued)

(continued)

Learning objectives (our translation)
77. Be able to recognize the importance of developing healthy habits
78. Realize personal responsibilities for one's own health, and actively participate in exercise, paying attention to personal care
79. Be able to sense and describe the features of objects, such as size, weight, shape, color, temperature etc.
80. Be able to make simple classifications and sort objects according to their characteristics
81. Be able to measure the common features of objects (length, weight, temperature) with simple instruments (length, balance, thermometer); Be able to design simple two-dimensional recording forms, make simple quantitative records with these forms, and use the appropriate units. On the basis of this experience, roughly measure other objects. Realize multiple measurements can improve the accuracy of measurement
82. Understand that the shape or size of an object can be changed by heating or cooling; and list some common effects of thermal expansion and contraction
83. Be able to judge that objects are composed of different materials, such as wood, metal, plastic and paper; and classify those objects according to their materials
84. Recognize some properties of materials (electrical conductivity, solubility, heat conduction, floating/sinking etc.); classify the materials according to those properties. Be able to link those properties of objects to their use/purpose
85. Be able to distinguish common natural materials and artificial materials. Realize that human beings are constantly inventing new materials in order to meet their needs. Enhance sensitivity towards new objects to stimulate awareness of innovation
86. Understand that there are three common states of matter: solid, liquid, and gas. Understand that changes in temperature can make materials change their states. Know the freezing point and boiling point of water
87. Know that there are two types of changes in matter, one is a mere change of form and the other will produce new substances
88. Understand that some of the changes of matter are reversible and others are irreversible. Recognize the impact of these changes on human lives
89. Know that some substances are renewable and others are non-renewable and recognize the importance of protecting resources
90. Realize that the use of materials has both favorable and detrimental aspects on human beings, and the importance of proper use of materials. Pay attention to safety and health; know some common measures for risk prevention, safety and health
91. Realize that the use of resources will bring both positive and negative effects on the environment, and human beings have a responsibility towards the environment
92. Be able to qualitatively describe the position of an object (front or rear, left or right, near or far, etc.); understand that to determine the position of an object it needs to be referenced with another object
93. Be able to measure and record the position of an object along a linear motion at different times; be able to show the relationship between the distance and time using simple charts or graphs
94. Know that to describe the motion of an object, we need to know the position, direction, and speed
95. Know that both pulling and pushing can change the movement of objects, pulling and pushing are both forces. Forces have both magnitude and direction

(continued)

(continued)

Learning objectives (our translation)
96. Know some common forces in daily life such as wind power, water power, gravity, elastic force, buoyancy and friction etc.
97. Investigate how to use a scale and lever to maintain balance
98. Know that using machines can improve work efficiency; understand how to use some simple machines such as ramps, levers, gears, pulleys etc.
99. Know that sound is produced by a vibrating object; be able to distinguish the magnitude and level of sounds
100. Know that sound needs to propagated through a material before reaching human ears
101. Be able to distinguish between musical tones and noise; understand the hazards of noise and how to prevent them
102. Know that heat energy can be transferred from one object to another
103. Know that temperature represents how hot or cold an object is; know the unit of temperature; and be able to use a thermometer
104. Understand that heat always flows from a hotter to a colder object until they are both at the same temperature. Understand some common methods of heat transfer and heat insulation
105. Understand the phenomenon that light travels in straight lines
106. Understand that mirrors or magnifying glasses can change the direction of light
107. Know that light is made up of colors; understand the phenomenon of chromatic dispersion of light
108. Recognize that electricity is a common energy source in life and work. Understand how to use electricity safely
109. Know that common electrical appliances require a complete (and closed) circuit to work; Know the function of the switch; Be able to make a simple circuit using some basic components
110. Know that some materials can readily conduct electricity, while others are unable to do so
111. Investigate the directional characteristics and poles of magnets; investigate the law that similar poles of magnets repel each other while unlike poles attract
112. Know that electricity can generate magnetism; investigate the factors which can influence the magnitude of electromagnetic force; understand the application of electromagnets
113. Know that equipment need energy to work, electricity, light, heat, sound, and magnetism are different forms of energy
114. Recognize that different forms of energy can transform into each other
115. Know the shape and size of the earth
116. Know that the surface composition of the earth is mainly water and less of that of land
117. Know that there is hot magma inside the earth
118. Understand the history of human thought regarding the shape of the earth
119. Understand the main legends and functions of globes and maps
120. Be able to classify rocks using different classification criteria
121. Know the names of the main energy producing minerals, metallic ores, and their extraction
122. Know the composition of soil
123. Be able to design experiments showing the effects of different soils on plant growth
124. Realize the close relationship between human survival and resources from the land and the importance of protecting those resources

(continued)

(continued)

Learning objectives (our translation)
125. Know the distribution of natural sources of water
126. Know that water can dissolve matter
127. Realize the close relationship between water and living things
128. Know the danger from and main reasons for water pollution
129. Be able to prove the existence of air using specific methods
130. Understand the use of the properties of air by Man
131. Know the significance of air to life. Experience the significance of air to life on earth and in relation to the effects of air on plants
132. Understand the adverse impact of human activities on the atmosphere, realize the importance of protecting the atmosphere
133. Know that the weather can be described with measurable indicators (such as temperature, wind direction, force of wind, precipitation, cloudiness, etc.)
134. Observe and collect data with a thermometer, a simple wind instrument, gauge; and be able to analyze these data and draw conclusions
135. Investigate the causes of rain and snow
136. Investigate the causes of wind. Observe an experiment showing that hot air rises and generates wind; discuss the causes of wind in the natural world
137. Realize that the long-term measurement and recording of weather data is useful
138. Be able to list examples showing the impact of weather changes on animal behavior
139. Know that the earth is constantly rotating, and that one rotation requires a day, 24 h
140. Understand the theories of early human societies concerning the causes of day and night and the contribution of Copernicus
141. Investigate the influence of day and night on the behavior of plants and animals
142. Understand that the earth's surface is constantly changing
143. Understand the phenomenon of volcanic eruptions
144. Understand the phenomenon of earthquakes
145. Recognize the role of various natural forces in changing the landscape
146. Understand the influence of human activities in changing the landscape
147. Recognize the influence of the changing seasons on plants and animals
148. Understand that the changing seasons are related to the earth's rotation around the sun
149. Know that the sun is a fireball with a very high temperature
150. Understand the use of solar energy by Man
151. Understand that there would be no life on earth without the sun
152. Know the sun's daily pattern of movement in the sky
153. Recognize that the changes of temperature and shadows in the day are related to the movement of the sun
154. Be able to identify directional position using the sun
155. Know that the moon is the Earth's satellite; know the moon's daily and monthly patterns of movement
156. Understand more about the moon from a variety of media
157. Know the composition of the solar system and the order of the nine planets

(continued)

(continued)

Learning objectives (our translation)
158. Know the representative constellations of the four seasons
159. Know the relationship between the solar system, galaxies, and the universe
160. Understand the history of human exploration of the universe
161. Know some important tools for exploring the universe; realize that human understanding of space has been deepened and enlarged with the progress of technology
162. Realize the arduous work that Man has done to explore the mysteries of the universe

Appendix B
Primary Science Learning Standards (Cognitive Domain) from Taiwan Based on Our English Translation

Section	Learning objectives (our translation)
	1. Topic of process skills
Observing	Grades 1–2
1-1-1-1	Use the five senses to observe the characteristics of objects (e.g., color, sound when tapped, odor, weight, etc.)
1-1-1-2	Be aware that some of properties of objects can undergo change and vary (e.g., ice melts when the temperature rises)
Comparing and classifying	Grades 1–2
1-1-2-1	Classify objects based on their features or properties (e.g., size, shading, etc.)
1-1-2-2	Compare patterns or objects to identify and communicate differences and similarities (e.g., a large and small tree although of different sizes are of the same type)
Organizing and connecting	Grades 1–2
1-1-3-1	Through a series of observations, describe how phenomena change (e.g., bean germination)
1-1-3-2	From a series of observations, comprehensively explain a significant phenomenon (e.g., a strong wind blows all the leaves on the ground, trees are blown down…)
Inducting and inferring	Grades 1–2
1-1-4-1	Be aware that events have causes, and believe that there are causal relationships
1-1-4-2	Be aware that with the same context and procedure, the results obtained should be similar or identical
Communicating	Grades 1–2
1-1-5-1	Learn how to use appropriate vocabulary to express observed phenomena (e.g., water can be described as boiling, hot, warm, cool or icy)
1-1-5-2	Attempt to understand the characteristics of objects from the description of others

(continued)

© The Author(s) 2017
Y.-J. Lee et al., *East-Asian Primary Science Curricula*,
SpringerBriefs in Education, DOI 10.1007/978-981-10-2690-4

63

(continued)

Section	Learning objectives (our translation)
Observing	Grades 3–4
1-2-1-1	Be aware that objects have identifiable characteristics and properties
Comparing and classifying	Grades 3–4
1-2-2-1	Use senses or ready-made tools to measure and do quantitative comparisons
1-2-2-2	Effectively use self-made criteria or self-made tools to do measurement
1-2-2-3	Understand that even with similar conditions, the results obtained may not be the same, and to be aware of the reasons for such outcomes
1-2-2-4	Know that if the goals (or attributes) are different, there are different ways of classification
Organizing and connecting	Grades 3–4
1-2-3-1	Describe the general features of data (eg for objects made of the same material, the greater the volume, the greater the weight…)
1-2-3-2	Be able to form predictions (eg that the ball can bounce high, because…)
1-2-3-3	Be able to control variables during experiments and make qualitative observations
Inducting and inferring	Grades 3–4
1-2-4-1	Using criteria to manage experimental data, derive conclusions
1-2-4-2	Use the results of experiments to explain or make predictions about phenomena
Communicating	Grades 3–4
1-2-5-1	Be able to use tables, graphs (such as interpretation of data and recording information)
1-2-5-3	Be able to obtain information from the telephone, newspapers, books, the Internet and media
Observing	Grades 5–6
1-3-1-1	Be able to perform procedures according to the experimental protocol
1-3-1-2	Be aware that a problem or phenomenon can be often observed from different angles and seen to have different characteristics
1-3-1-3	Distinguish the difference between original amount and the rate of change (such as temperature and temperature changes)
Comparing and classifying	Grades 5–6

(continued)

(continued)

Section	Learning objectives (our translation)
1-3-2-1	Before an experiment, estimate possible changes in a "variable" in terms of amount and scope
1-3-2-2	From the amount of change and the volume change, assess the degree of change
1-3-2-3	Depending on the degree of differences, perform classifications beyond the second level
Organizing and connecting	Grades 5–6
1-3-3-1	During experiments, confirm the causes of change and perform manipulations
1-3-3-2	Find relationships between the dependent variable and independent variable
1-3-3-3	From a series of related activities, describe the main features of these activities
Inducting, judging and inferring	Grades 5–6
1-3-4-1	Organize a coherent explanation from a number of different sources of information
1-3-4-2	Identify the characteristics of data and interpret their general aspects
1-3-4-3	From displays of relevant information, suggest a possible causal relationship
1-3-4-4	Evaluate and judge the outcomes of experiments
Communicating	Grades 5–6
1-3-5-1	Use appropriate graphs to express data
1-3-5-2	Express data in an appropriate manner (e.g. the number of lines, tables, graphs)
1-3-5-3	Describe clearly the process of scientific inquiry and outcomes
1-3-5-5	Listen to other people's reports and make appropriate responses
	2. Topic of knowledge of science and technology
Knowledge level	Grades 1–2
2-1-1-1	Use the five senses to observe natural phenomena, recognise various phenomena in their natural state and state the changes. Use appropriate vocabulary to report and describe observations. Use of ready-made tables, charts, to express the observed data
2-1-1-2	Be aware that changes of state are often caused by a number of reasons, and practice how to plan and conduct inquiry (here)
Recognize common animals and plants	Grades 1–2
2-1-2-1	Select one (or a certain type of) plant and animal, observe continuously and learn how to record any major changes Be aware that plants will grow, be aware that plants have distinguishing features for identification. Note that soil is

(continued)

(continued)

Section	Learning objectives (our translation)
	required for plant growth as well as sunlight and moisture in the environment. Be aware of how animals feed, what they eat, what they do, and changes in body shape occur during growth
Observe phenomena and their changes	Grades 1–2
2-1-3-1	Observe changes in phenomena (e.g., changes in the weather, states of object), be aware that changes in phenomenon must have reasons
2-1-3-2	Construct a variety of different toys, experience different kinds of "forces" and how forces can move objects, or how vibrating objects produce sounds
2-1-4-1	Recognize and use everyday technology (media, transport, safety equipment)
Knowledge level	Grades 3–4
2-2-1-1	Be aware that natural phenomena have causes. Be able to use tools such as a thermometer, magnifying glass, mirrors to help make observations, and conduct of inquiry based on these changes, and learn how to organize observations and work methods
Understand growth of animals and plants	Grades 3–4
2-2-2-1	Grow a plant, keep a small animal, and share about these experiences. Know the requirements for growing plants, learn how to provide sunlight, moisture, fertilizers, and select soil cultivation techniques
2-2-2-2	Understand the characteristic appearances of terrestrial (or aquatic) animals, their movement, and note how to improve their living environment, adjust their diets to maintain their health
Awareness of materials	Grades 3–4
2-2-3-1	Recognize that besides the physical features of objects, they also have different properties such as solubility, magnetism, electrical conductivity, etc. And apply these properties to separate or combine them. Aware that materials may be burnt, oxidized, fermented and undergo other changes due to temperature, water, air
2-2-3-2	Recognize the importance of water and its properties
Awareness of the environment	Grades 3–4
2-2-4-1	Understand how to use air temperature, wind direction, wind speed, rainfall to describe the weather. Discover that the weather will change, be aware that the amount of water vapor plays a very important role in the change in the weather

(continued)

(continued)

Section	Learning objectives (our translation)
2-2-4-2	Observe that the moon rises from the east and sets in the west, and make long-term continuous observation of the phase of the moon, discover that moon phases have periodicity
Recognize uses of interactions	Grades 3–4
2-2-5-1	Use refraction, dispersion, batteries, wires, bulbs, small motors, air or water flow to design a variety of toys. Look for ways to improve toys, discuss the reasons for the change, and gain understanding into the nature of the material, and then to begin to learn about their improvement
Recognize common technologies	Grades 3–4
2-2-6-1	Understand media equipment, such as recording and video equipment
2-2-6-2	Recognize energy used for transport (such as gasoline) and transport equipment (such as locomotives, cars, tractors)
Knowledge level	Grades 5–6
2-3-1-1	Ask questions, learn problem solving strategies, study changes, observe changes and speculate causal relationships. Learn how to manage data, design tables and graphs to represent data. Learn about the relationship between independent and dependent variable, hypothesize or make a appropriate explanation
Recognize the ecology of plants and animals	Grades 5–6
2-3-2-1	Be aware of the features of plant roots, stems, leaves, flowers, fruits, and seeds. Illumination, temperature, humidity, soil influence the growth of plants. Plants vary due to adaptation to different habitats. Discover that there are many methods for plant propagation
2-3-2-2	Observe the morphology and movement of animals and their differences and similarities. Observe how animals maintain body temperature, feeding, reproduction, communicate, engage in social behavior and adapt to their habitat as part of animal ecology and life
2-3-2-3	Understand that animals have oviparous and viviparous reproductive modes. Discover that there are similarities between the parents and offspring of animals and plants as well as differences
2-3-2-4	With the knowledge of animals and plants, customize some methods of animal and plant classification
Recognize materials	Grades 5–6
2-3-3-1	Understand the properties of matter, investigate how light, temperature, and air affects the properties of matter

(continued)

(continued)

Section	Learning objectives (our translation)
2-3-3-2	Investigate the properties of oxygen and carbon dioxide; the production of oxygen, understand combustion, oxidation (rust), the production of carbon dioxide and its features (dissolving in water), air pollution and other phenomena
2-3-3-3	Investigate the properties of dissolved materials, water conductivity, pH, evaporation, diffusion, expansion and contraction, and hard and soft water etc.
2-3-3-4	Recognize environments that promote the oxidation reaction
Awareness of the environment	Grades 5–6
2-3-4-1	Through long-term observation, discover changes in sun rise and the azimuth (or maximum elevation angle), at the same time at night, the constellations across the seasons are different, but they change regularly year by year
2-3-4-2	Recognize high and low pressure lines, fronts on the weather map. Observe (through data collection) the start and end of a typhoon
2-3-4-3	Know that high and low temperatures can influence the different states of water, which also causes the formation of frost, dew, clouds, rain, and snow
2-3-4-4	Know that in the living environment, air, earth and water all interact with each other
Recognize uses of interactions	Grades 5–6
2-3-5-1	Know that heat travels from a hotter to colder region, transmission methods include conduction, convection, radiation. Transmission would be different according to different materials and the nature of spaces. This knowledge can be applied to the insulation or removal of heat
2-3-5-2	Understand by making the instruments the causes of quality of sound, volume, timbre, etc. know the difference between tone and noise
2-3-5-3	Understand that the size of a force can be measured from the amount of deformation or motion of objects
2-3-5-4	Through the use of simple machines, understand that force can be used to move levers, belts, gears, fluid (pressure) and others
2-3-5-5	Understand that a current can generate a magnetic field. Make an electromagnet, understand geomagnetism and the compass. Discover that some "forces" need not have points of contact to work, such as gravity, magnetism
Recognize common technologies	Grades 5–6
2-3-6-1	Recognize the raw materials for manufacturing everyday objects (such as wood, metal, plastic)
2-3-6-2	Recognize the structure and materials for housing
2-3-6-3	Recognize information technology equipment
	3. Topic of nature of science and technology

(continued)

(continued)

Section	Learning objectives (our translation)
	Grades 1–2
3-1-0-1	Able to describe observed phenomenon in one's own words
	Grades 3–4
3-2-0-1	Know that verification or testing methods can be used to check one's thinking
3-2-0-2	Be aware that if experimental conditions are the same, the outcomes will be very similar
	Grades 5–6
3-3-0-1	Be able to conduct scientific inquiry activities, and understand that scientific knowledge has been verified
3-3-0-2	Know that some phenomena (such as UFO) have difficulties of evidence collection, and cannot undergo scientific experimentation
3-3-0-3	Discover that using scientific knowledge to make inferences, confirmation of particular phenomena is possible
3-3-0-4	Be aware that "examining existing data from a new perspective" or "examining new data from existing theories", can often generate new questions
3-3-0-5	Be aware that sometimes under the same experimental conditions, there may be uncontrolled factors, thus making the outcomes different
	4. Topic of development of science and technology
Nature of S and T	Grades 3–4
4-2-1-1	Understand the importance of science and technology in daily life
4-2-1-2	Recognize the characteristics of science and technology
S and T and the environment	Grades 3–4
4-2-2-2	Recognize common household products
Nature of S and T	Grades 5–6
4-3-1-1	Recognize different aspects of science and technology
4-3-1-2	Understand equipment, materials and energy
Evolution of S and T	Grades 5–6
4-3-2-1	Recognize the science of the agricultural era
4-3-2-2	Recognize the science of industrialization era
4-3-2-3	Recognize the science of information era
4-3-2-4	Recognize domestic and external technological inventions
S and T and the environment	Grades 5–6
4-3-3-1	Understand common local transport facilities, recreational facilities and other technologies
	5. Topic of habits of thinking
Problem solving	Grades 1–2
6-1-2-2	Learn the organization of work procedures
6-1-2-3	Learn how to allocate work, how to collaborate with others to complete a project

(continued)

(continued)

Section	Learning objectives (our translation)
Critical thinking	Grades 3–4
6-2-1-1	Be able to asking various questions such as "What is this?", "How could this be?" and raise researchable questions
6-2-2-1	Be able to frequently ask oneself "How to do it?" to prepare oneself to think about a solution
Critical thinking	Grades 5–6
6-3-1-1	From the data or reports of others, raise appropriate confirmations and questions
Idea generation	Grades 5–6
6-3-2-1	Observe that different approaches can often achieve the same outcomes
6-3-2-3	When facing problems, be able to think of alternative ideas, and propose solutions
Problem solving	Grades 5–6
6-3-3-1	Be able to plan, organize inquiry activities
	6. Topic of applications of science
	Grades 1–2
7-1-0-1	Learn how to organize work, be methodical
7-1-0-2	Learn how to operate a variety of simple equipment
	Grades 3–4
7-2-0-1	Use scientific knowledge to solve problems (e.g., knowing the temperature level to consider dressing warmly)
7-2-0-2	Be able to use the spirit and methods of scientific inquiry to perform tasks
7-2-0-3	Be able to safely and properly use everyday appliances
	Grades 5–6
7-3-0-1	Be aware that experimental or scientific knowledge can predict possible events
7-3-0-3	Be able to plan, organize inquiry activities
7-3-0-4	Be aware that many ingenious instruments are often the simple application of scientific principles
	8. Topic of design and production
	Grades 5–6
8-3-0-1	Be able to use association, brainstorming, concept maps and other procedures and the development of creative ideas to express one's decisions for product revision
8-3-0-2	Use a variety of thinking methods, ponder how objects can change function and form
8-3-0-3	Recognize and design basic shapes
8-3-0-4	Understand the prototyping process

A-B-C-D nomenclature: "A" indicates the main topic (8 in total eg 1 = process skills), "B" indicates the grade level (1 for Grades 1–2, 2 for 3–4, 3 for 5–6), "C" represents the number of the sub-topic in the syllabus and "D" shows the item number in the category

Appendix C
Primary Science Learning Objectives (Cognitive Domain) from Korea Based on Our English Translation

Items	Learning objectives (our translation)
Grade 3–4 1. Characteristics of materials and substances	
4sci 01-01	Make connections between the functions and characteristics of objects by comparing objects made of different materials and substances
4sci 01-02	Compare various characteristics of materials and substances by observing objects with the same shapes and sizes but of different materials and substances
4sci 01-03	Explain the changes in the characteristics of substances by observing the differences before and after different substances are mixed
4sci 01-04	Design various objects by choosing various materials and substances and discuss their strengths and weakness
Grade 3–4 2. Usage of magnets	
4sci 02-01	Identify the two poles of magnets by observing the forces of attraction and repulsion between magnets
4sci 02-02	Explain the unidirectional orientation of compass needles through observations
4sci 02-03	Investigate examples of the use of magnets in everyday lives and explain the functions of magnets in relation to their characteristics
Grade 3–4 3. The lives of animals	
4sci 03-01	Classify animals according to their characteristics through observing various animals
4sci 03-02	Explain how the appearance and behavior of animals are related to their habitats
4sci 03-03	Make a presentation of everyday examples of man-made objects that resemble the characteristics of animals
Grade 3–4 4. Changes of the earth surface	
4sci 04-01	Observe soils from various places and compare them
4sci 04-02	Explain how soils are produced using models
4sci 04-03	Relate the characteristics of geographical features of rivers and the ocean with the functions of streams and the ocean
Grade 3–4 5. Lives of plants	
4sci 05-01	Classify plants according to their characteristics through observing various plants
4sci 05-02	Explain how the appearance and life cycles of plants are related to their habitats
4sci 05-03	Make a presentation of everyday examples of man-made objects that resemble the characteristics of plants

(continued)

© The Author(s) 2017
Y.-J. Lee et al., *East-Asian Primary Science Curricula*,
SpringerBriefs in Education, DOI 10.1007/978-981-10-2690-4

(continued)

Items	Learning objectives (our translation)
Grade 3–4 6. Strata and fossils	
4sci 06-01	Observe various strata and explain how strata are formed using models
4sci 06-02	Classify sedimentary rocks according to the size of particles and explain how sedimentary rocks are made using models
4sci 06-03	Understand how fossils are formed and infer about prehistoric life and the environment through observation of fossils
Grade 3–4 7. States of substance	
4sci 07-01	Explain the characteristics of solids and liquids by observing their changes of shape and volume in different containers
4sci 07-02	Conduct an experiment that shows air occupies space
4sci 07-03	Conduct an experiment that shows air has mass
4sci 07-04	Classify everyday materials into solids, liquids, and gases
Grade 3–4 8. Characteristics of sound	
4sci 08-01	Explain objects that make sound vibrate through observation of various objects that make sound
4sci 08-02	Compare the loudness and pitch of sound
4sci 08-03	Observe how sound moves through and reflects off various objects and discuss ways of reducing noise
Grade 3–4 9. Weight	
4sci 09-01	Investigate examples of measuring the weight of objects in everyday lives and explain reasons why we need to measure weight
4sci 09-02	Compare the weight of objects through balancing activities
4sci 09-03	Investigate the relationship between the length of spring extension and a weight and explain the principles of measuring weight
4sci 09-04	Design and build a simple scale and test it
Grade 3–4 10. Lifecycles of animals	
4sci 10-01	Compare the characteristics of male and female animals and explain the different roles of male and female during reproduction
4sci 10-02	Plan observations of the lifecycles of animals, observe an animal by raising it, and express observations through writing and drawing
4sci 10-03	Investigate the lifecycles of various animals and explain that there are different types of animal lifecycles
Grade 3–4 11. Volcanoes and earthquakes	
4sci 11-01	Explain the various substances produced from volcanic activities
4sci 11-02	Understand how igneous rocks are produced and compare characteristics of granite and basalt
4sci 11-03	Make a presentation on the impact of volcanic activities on everyday lives
4sci 11-04	Understand the causes of earthquakes and discuss methods of earthquake safety
Grade 3–4 12. Separation of mixtures	
4sci 12-01	Find examples of mixtures in everyday lives and explain the necessities of separating mixtures
4sci 12-02	Separate solid mixtures by using the size of particles and magnetism

(continued)

(continued)

Items	Learning objectives (our translation)
4sci 12-03	Separate mixtures of soluble and insoluble materials by filtration
4sci 12-04	Separate solids that have been dissolved in water through evaporation
Grade 3–4 13. Lifecycles of plants	
4sci 13-01	Explain the necessary conditions of germination and growth
4sci 13-02	Plan a way to observe the lifecycles of plants and observe the lifecycles by growing plants
4sci 13-03	Investigate the lifecycles of various plants and explain that there are different types of plant lifecycles
Grade 3–4 14. Changes of states of water	
4sci 14-01	Know that water can transform into water vapor or ice and observe the changes in volume and weight when water freezes and ice thaws
4sci 14-02	Know the changes when water evaporates and boils through observation and find everyday examples related to this phenomena
4sci 14-03	Observe the phenomenon of condensation of water vapor and find everyday examples related to this
Grade 3–4 15. Shadows and mirrors	
4sci 15-01	Explain how shadows are formed by observing the shadows of various objects
4sci 15-02	Observe and describe the changes in the size of shadows by varying the distance between an object and a light source
4sci 15-03	By comparing an object and its reflection in mirrors, explain the characteristics of mirrors
4sci 15-04	Investigate examples of using mirrors in everyday lives and explain the function of mirrors in relation to the characteristics of mirrors
Grade 3–4 16. The shape of the earth	
4sci 16-01	Explain the shape of the earth and its surface through investigating earth-related resources
4sci 16-02	Explain the characteristics of the ocean by comparing it with the land
4sci 16-03	Explain the roles of air surrounding the earth by giving examples
4sci 16-04	Investigate the earth and the moon, understand the shape, surface, and environments and compare the earth and the moon
Grade 3–4 17. The travel of water (integrated topic)	
4sci 17-01	Explain the water cycle and state changes in relation to various phenomena among living beings, the earth, and the air
4sci 17-02	Understand the importance of water and investigate creative ways to solve water shortages
Grade 5–6 1. Temperature and heat	
6sci 01-01	Investigate examples of measuring and estimating temperature in everyday lives and explain reasons for accurate measurement of temperature
6sci 01-02	Observe that the different temperatures of two objects will become the same after contact and explain the changes of temperature in terms of heat transfer
6sci 01-03	Compare the rate of heat transfer in different kinds of solids through observations and investigate examples of heat insulation in everyday lives

(continued)

(continued)

Items	Learning objectives (our translation)
6sci 01-04	Observe convection in liquids and gases and explain heat transfer in convection
Grade 5–6 2. The solar system and stars	
6sci 02-01	Understand that the sun is the source of energy of the earth and investigate the sun and planets in the solar system
6sci 02-02	Know what stars are and investigate the major constellations
6sci 02-03	Find the North Star by using constellations in the Northern sky
Grade 5–6 3. Dissolution and solution	
6sci 03-01	Observe the phenomenon of a substance dissolving in water and explain what are solutions
6sci 03-02	Compare the amount of solute dissolving in water which is different depending on the types of solute
6sci 03-03	Conduct an experiment to show that the amount of solute dissolving differs depending on the temperature of water
6sci 03-04	Design ways to compare the relative concentrations of solutions
Grade 5–6 4. Various living beings and our lives	
6sci 04-01	Explain the types and characteristics of living beings through investigating other living beings other than animals and plants
6sci 04-02	Discuss positive and negative impacts that various living beings have on our everyday lives
6sci 04-03	Investigate and present examples of using advanced life sciences in everyday lives
Grade 5–6 5. Life and the environment	
6sci 05-01	Know that an ecosystem consists of living and non-living objects and explain how the components of ecosystems impact each other
6sci 05-02	By understanding the impact of non-living objects on living beings, explain the relationships between the environment and living beings
6sci 05-03	Perceive the importance of conserving the ecosystem and discuss our roles and actions for it
Grade 5–6 6. Weather and our lives	
6sci 06-01	Measure humidity and investigate examples of the impact of humidity in everyday lives
6sci 06-02	Understand the similarities and differences among dew, fog, and clouds and explain the formation of rain and snow
6sci 06-03	Know what are high and low pressure and explain the causes of wind
6sci 06-04	Relate the characteristics of seasonal weather to the characteristics of surrounding air around Korea
Grade 5–6 7. Movements of objects	
6sci 71-01	By observing the movements of objects in everyday lives, compare speed qualitatively
6sci 07-02	By investigating the distance and time of an object's movement, calculate velocity
6sci 07-03	Investigate and present examples of safety and safety devices on speed in everyday lives

(continued)

(continued)

Items	Learning objectives (our translation)
Grade 5–6 8. Acids and Bases	
6sci 08-01	Classify various solutions in everyday lives with various criteria
6sci 08-02	Classify various solutions into acids and bases by using pH indicators
6sci 08-03	Compare the characteristics of acids and bases and observe changes when acids and bases are mixed
6sci 08-04	Investigate and present examples of using acids and bases in everyday lives
Grade 5–6 9. Movements of the earth and moon	
6sci 09-01	Explain how the location of the sun and the moon changes throughout the day in terms of the rotation of the earth
6sci 09-02	Explain how constellations change according to the seasons in terms of the revolution of the earth
6sci 09-03	Observe how the shape and location of the moon changes periodically
Grade 5–6 10. Various gases	
6sci 10-01	Produce oxygen and carbon dioxide with experimental apparatus and check and explain the characteristics of these gases
6sci 10-02	Observe the phenomena that the volume of gas changes according to temperature and pressure and find examples in everyday lives
6sci 10-03	Investigate and present various gases that comprise the air
Grade 5–6 11. Light and lenses	
6sci 11-01	By observing various colors of sunlight through a prism, explain that sunlight consists of various colors of light
6sci 11-02	By observing the phenomena of light refraction through glass, water, convex lenses, express observations through drawings
6sci 11-03	Observe objects through convex lenses and investigate the use of convex lenses
Grade 5–6 12. Structures and functions of plants	
6sci 12-01	Observe cells which are the basic units of living organisms using microscopes
6sci 12-02	By observing the whole structure of plants and doing experiments, explain the structure and function of roots, stems, leaves, and flowers
6sci 12-03	Investigate the methods of seed dispersal of various plants and explain various ways of seed dispersal
Grade 5–6 13. Usage of electricity	
6sci 13-01	By connecting batteries, bulbs, and wires, explain the necessary conditions to light bulbs
6sci 13-02	Compare the brightness of bulbs in series and parallel circuits
6sci 13-03	Discuss ways of saving electricity and how to use it safely
6sci 13-04	By making electromagnets, compare magnets and electromagnets and investigate examples of the use of electromagnets in everyday lives
Grade 5–6 14. Seasonal changes	
6sci 14-01	By measuring the sun's altitude, length of shadow, and temperature throughout the day, find relationships among them
6sci 14-02	Explain the changes of the sun's altitude, the length of day and night, and temperature change throughout the seasons

(continued)

(continued)

Items	Learning objectives (our translation)
6sci 14-03	Explain through model-based experiments that seasons change because the earth revolves around the sun with the rotation axis tilted
Grade 5–6 15. Combustion and fire extinction	
6sci 15-01	Observe the common phenomena of combustion and find the conditions for combustion
6sci 15-02	Through experiments, find substances produced by combustion
6sci 15-03	Suggest ways to extinguish fire in relation to the conditions of combustion and discuss fire safety rules
Grade 5–6 16. Structures and functions of our bodies	
6sci 16-01	By understanding the features and functions of bones and muscles, explain how bodies move
6sci 16-02	Explain the types, location, features and functions of the digestive, circulatory, respiratory, and excretory systems
6sci 16-03	Know the types, locations, features, and functions of the sensory system and explain how stimuli are transmitted
6sci 16-04	By observing the changes in our bodies when exercising, explain how various systems are interrelated
Grade 5–6 17. Energy and living (integrated topic)	
6sci 17-01	Know that living beings and machines need energy to survive and move and investigate what types of energy are used in that process
6sci 17-02	Know that energy transforms into different types through examples of natural phenomena and everyday lives and discuss the ways of using energy in efficient ways

Appendix D
Primary Science Learning Objectives (Cognitive Domain) from Singapore [The Government of Singapore (c/o Ministry of Education) Owns the Copyright to This Table and the Table has been Reproduced with Their Permission]

Theme	Learning objectives
Diversity	*Recognise some broad groups of living things – plants (flowering, non-flowering) – animals (amphibians, birds, fish, insects, mammals, reptiles) – fungi (mould, mushroom, yeast) – bacteria
Diversity	*Describe the characteristics of living things – need water, food and air to survive – grow, respond and reproduce
Diversity	*Observe a variety of living and non-living things and infer differences between them
Diversity	*Classify living things into broad groups (in plants and animals) based on similarities and differences of common observable characteristics
Diversity	*Relate the use of various types of materials (ceramic, fabric, glass, metal, plastics, rubber, wood) to their physical properties
Diversity	*Compare physical properties of materials based on: – strength – flexibility – waterproof – transparency – ability to float/sink in water
Cycles	**Recognise the importance of water to life processes
Cycles	**Recognise processes in the sexual reproduction of flowering plants – pollination – fertilisation (seed production) – seed dispersal – germination
Cycles	**Recognise the similarity in terms of fertilisation in the sexual reproduction of flowering plants and humans
Cycles	**Recognise the process of fertilisation in the sexual reproduction of humans

(continued)

© The Author(s) 2017
Y.-J. Lee et al., *East-Asian Primary Science Curricula*,
SpringerBriefs in Education, DOI 10.1007/978-981-10-2690-4

(continued)

Theme	Learning objectives
Cycles	*Show an understanding that different living things have different life cycles – Plants – Animals
Cycles	**Show an understanding that living things reproduce to ensure continuity of their kind and that many characteristics of an organism are passed on from parents to offspring
Cycles	*Observe and compare the life cycles of plants grown from seeds over a period of time
Cycles	*Observe and compare the life cycles of animals over a period of time (butterfly, beetle, mosquito, grasshopper, cockroach, chicken, frog)
Cycles	**Investigate the various ways in which plants reproduce and communicate findings – spores – seeds
Cycles	**Describe the impact of water pollution on Earth's water resources
Cycles	*State that matter is anything that has mass and occupies space
Cycles	**Recognise that water can exist in three interchangeable states of matter
Cycles	**Recognise the importance of the water cycle
Cycles	**Show an understanding of how water changes from one state to another – Melting (solid to liquid) – Evaporation/Boiling (liquid to gas) – Condensation (gas to liquid) – Freezing (liquid to solid)
Cycles	*Differentiate between the three states of matter (solid, liquid, gas) in terms of shape and volume
Cycles	**Compare water in 3 states
Cycles	**Show an understanding of the terms melting point of ice (or freezing point of water) and boiling point of water
Cycles	**Show an understanding of the roles of evaporation and condensation in the water cycle
Cycles	*Measure mass and volume using appropriate apparatus
Cycles	**Investigate the effect of heat gain or loss on the temperature and state of water and communicate findings – when ice is heated, it melts and changes to water at 0 °C – when water is cooled, it freezes and changes to ice at 0 °C – when water is heated, it boils and changes to steam at 100 °C – when steam is cooled, it condenses to water
Cycles	**Investigate the factors which affect the rate of evaporation and communicate findings – wind – temperature – exposed surface area
Systems	**Identify electrical conductors and insulators
Systems	**Recognise that an electric circuit consisting of an energy source (battery) and other circuit components (wire, bulb, switch) forms an electrical system
Systems	**Show an understanding that a current can only flow through a closed circuit
Systems	**Construct simple circuits from circuit diagrams

(continued)

(continued)

Theme	Learning objectives
Systems	**Investigate the effect of some variables on the current in a circuit and communicate findings – number of batteries (arranged in series) – number of bulb (arranged in series and parallel)
Systems	**Show an understanding that a cell is a basic unit of life
Systems	*Identify the organ systems and state their functions in human (digestive, respiratory, circulatory, skeletal and muscular)
Systems	*Identify the organs in the human digestive system (mouth, gullet, stomach, small intestine and large intestine) and describe their functions
Systems	**Identify the organs of the human respiratory and circulatory systems and describe their functions
Systems	**Recognise the integration of the different systems (digestive, respiratory and circulatory) in carrying out life processes
Systems	*Identify the different parts of plants and state their functions – leaf – stem – root
Systems	**Identify the parts of the plant transport system and describe their functions
Systems	**Identify the different parts of a typical plant cell and animal cell and relate the parts to the functions – parts of plant cell: cell wall, cell membrane, cytoplasm, nucleus and chloroplasts – parts of animal cell: cell membrane, cytoplasm, nucleus
Systems	*Observe plant parts
Systems	**Compare how plants, fish and humans take in oxygen and give out carbon dioxide
Systems	**Compare the ways in which substances are transported within plants and humans – plants: tubes that transport food and water – humans: blood vessels that transport digested food, oxygen and carbon dioxide
Systems	**Investigate the functions of plant parts and communicate findings – leaf – stem – root
Systems	**Compare a typical plant and animal cell
Systems	**Recognise that air is a mixture of gases such as nitrogen, carbon dioxide, oxygen and water vapour
Interactions	*List some uses of magnets in everyday objects
Interactions	*Recognise that a magnet can exert a push or a pull
Interactions	*Identify the characteristics of magnets – magnets can be made of iron or steel – magnets have two poles. A freely suspended bar magnet comes to rest pointing in a North-South direction – unlike poles attract and like poles repel – magnets attract magnetic materials

(continued)

(continued)

Theme	Learning objectives
Interactions	*Compare magnets, non-magnets and magnetic materials
Interactions	*Make a magnet by the "Stroke" method and the electrical method
Interactions	**Recognise that adaptations serve to enhance survival and can be structural or behavioural – cope with physical factors – obtain food – escape predators – reproduce by finding and attracting mates or dispersing seeds/fruits
Interactions	**Identify the factors that affect the survival of an organism – physical characteristics of the environment (temperature, light, water) – availability of food – types of other organisms present (producers, consumers, decomposers)
Interactions	**Differentiate among the terms organism, population and community – An organism is a living thing – A population is defined as a group of plants and animals of the same kind, living and reproducing at a given place and time – A community consists of many populations living together in a particular place
Interactions	**Show an understanding that different habitats support different communities (garden, field, pond, seashore, tree, mangrove swamp)
Interactions	**Give examples of man's impact, (both positive and negative) on the environment
Interactions	**Discuss the effect on organisms when the environment becomes unfavourable (organisms adapt and survive; move to other places or die)
Interactions	**Trace the energy pathway from the Sun through living things and identify the roles of various organisms (producers, consumers, predators, prey) in a food chain and a food web
Interactions	**Observe, collect and record information regarding the interacting factors within an environment
Interactions	**Identify a force as a push or a pull
Interactions	**Show an understanding of the effects of a force – A force can move a stationary object – A force can speed up, slow down or change the direction of motion – A force can stop a moving object – A force may change the shape of an object
Interactions	**Recognise and give examples of the different types of forces – magnetic force – gravitational force – elastic spring force – frictional force
Interactions	**Investigate the effect of friction on the motion of objects and communicate findings
Interactions	**Recognise that objects have weight because of the gravitational force acting on the object
Interactions	**Investigate the effects of forces on springs and communicate findings
Energy	*Recognise that an object can be seen when it reflects light or when it is a source of light

(continued)

(continued)

Theme	Learning objectives
Energy	*Recognise that a shadow is formed when light is completely or partially blocked by an object
Energy	*Investigate the variables that affect shadows formed and communicate findings – shape, size and position of object(s) – distance between light source-object and object-screen
Energy	*List some common sources of heat
Energy	*State that the temperature of an object is a measurement of its degree of hotness
Energy	*List some effects of heat gain/loss in our everyday life – contraction/expansion of objects (solid, liquid and gas) – change in state of matter
Energy	*Identify good and bad conductors of heat – good conductors: metals – poor conductors: wood, plastics, air
Energy	*Differentiate between heat and temperature – heat is a form of energy – temperature is a measurement of the degree of hotness of an object
Energy	*Show an understanding that heat flows from a hotter to a colder object/region/place until both reach the same temperature
Energy	*Relate the change in temperature of an object to the gain or loss of heat by the object
Energy	*Measure temperature using a thermometer and a datalogger with temperature/heat sensors
Energy	**Recognise that the Sun is our primary source of energy (light and heat)
Energy	**Recognise that energy from most of our energy resources is derived in some ways from the Sun
Energy	**State that living things need energy to carry out life processes
Energy	**Differentiate the ways in which plants and animals obtain energy
Energy	**Investigate the requirements (water, light energy and carbon dioxide) for photosynthesis (production of sugar and oxygen) and communicate findings
Energy	**Recognise and give examples of the various forms of energy – kinetic energy – potential energy – light energy – electrical energy – sound energy – heat energy
Energy	**Investigate energy conversion from one form to another and communicate findings

*Grades 3–4
**Grades 5–6

Printed in the United States
By Bookmasters